SPECIFIC PRODUCTION OF AQUATIC INVERTEBRATES

SPECIFIC PRODUCTION OF AQUATIC INVERTEBRATES

V. E. Zaika

Translated from Russian by A. Mercado
Translation edited by B. Gollek

A HALSTED PRESS BOOK

JOHN WILEY & SONS
New York · Toronto

ISRAEL PROGRAM FOR SCIENTIFIC TRANSLATIONS
Jerusalem · London

© 1973 Israel Program for Scientific Translations Ltd.

Sole distributors for the Western Hemisphere and Japan

HALSTED PRESS, a division of
JOHN WILEY & SONS, INC., NEW YORK

Library of Congress Cataloging in Publication Data

Zaika, Viktor Evgen'evich.

    Specific production of aquatic invertebrates.

    "A Halsted Press book."

    Translation of *Udel'naĩa produkt͡siĩa vodnykh bespozvonochnykh.*

    1. Aquatic invertebrates.   2. Biological productivity. I. Title.

QL120.Z3413   592'.05'263   73-12320
ISBN 0-470-98111-3

Distributors for the U. K., Europe, Africa and
the Middle East

JOHN WILEY & SONS, LTD., CHICHESTER

Distributed in the rest of the world by

KETER PUBLISHING HOUSE JERUSALEM LTD.

ISBN 0 7065 1352 5
IPST cat. no. 22077

This book is a translation from Russian of
UDEL'NAYA PRODUKTSIYA VODNYKH
BESPOZVONOCHNYKH
Izdatel'stvo "Naukova Dumka"
Kiev, 1972

Printed in Israel

Contents

INTRODUCTION

The problem of animal productivity, studied up to now mainly by hydrobiologists, is far from new. It should be noted that assessments of plankton production were already made by Hensen (1887). Boysen-Jensen (1919) strictly defined the term "production" and gave a method for its determination. However, later publications on the productivity of populations and communities often presented vague and wide interpretations of the terms "production" and "productivity," causing considerable confusion and some measure of distrust. It was proposed to abandon the terms "production" and "productivity" and to discontinue research on these topics. Soviet hydrobiologists joined the general trend toward a theoretical and generalized approach to the problem, although the original goals of research on productivity had been formulated quite concisely and concretely (Zenkevich, 1931, 1934; Brotskaya and Zenkevich, 1936).

Discussions on problems of productivity have continued for several decades; many reviews of different positions taken by biologists have been published (Karzinkin, 1952; Vodyanitskii, 1954; MacFadyen, 1963; Davis, 1963; etc.). Particular attention was devoted to studies of the density and biomass of animals, whereas production itself and *P/B* coefficients were largely ignored.

Owing to the scarcity of factual material, the predominantly theoretical approach to this problem has gradually been replaced by increasingly accurate production calculations during the last decade. The transition to production calculations was accompanied by an extensive search for calculation methods. Errors did occur, but several correct and sufficiently accurate methods for calculating production are available today (see Vinberg, 1968 a). The actual work of measuring or assessing various parameters rapidly replaced diverse and vague definitions of basic terms; the best conception of a given parameter is obtained by determining it practically. Production, as it is understood today, corresponds exactly to the definition by Boysen-Jensen (1919).

As a result, data on the production of various species of aquatic fauna (mainly crustaceans) are rapidly accumulating. With the appearance of the detailed, thorough handbook mentioned above, this work continues. At the same time it is necessary to review

1

results from the theoretical viewpoint and to select basic directions for further research.

The study of productivity, naturally, cannot be reduced to mere calculation of the productions of different populations and communities. The material collected should serve as the basis for generalizations and for establishing quantitative relations between productivity and other characteristics of the biosystems. Knowledge of the principles determining the productivities of biosystems may make it possible to forecast productivity under specific conditions, according to given characteristics of the system or the environment. This is all the more important in connection with the increasing intervention of man in natural processes. In the strict sense the study of productivity includes determination of the production, specific production and, to some extent, the biomass of the system. In most cases, however, the problem of productivity is treated far more generally as a major ecological issue which covers a wide array of topics. Such an approach emphasizes theoretical analysis of the interrelations between various ecological problems.

Having defined the analysis of principles determining the productivity level of aquatic invertebrate populations as our principal task, it was necessary to examine a large number of problems. First, existing concepts of the production process in biosystems had to be reviewed and the most effective comparative index was selected (specific production was found most suitable for this purpose). Furthermore, it was necessary to collect and classify available data on the specific productions of different aquatic invertebrates. This involved a critical analysis of current methods of calculating production parameters, and discussions on some poorly studied methodical problems. Using published data on growth and age structure of different animal populations, initial assessments were made of the specific productions of many species. Biological data necessary for assessment of production were obtained for a number of Black Sea invertebrates. The factual basis for the study of patterns of quantitative relations between the production of the population and such indexes as age structure of the population, growth rate, individual life span and temperature of the environment was provided by the specific productions of the different animals. Mathematical models were widely used in the analysis of these problems.

Since the main purpose of this work was to analyze the quantity of the specific productions of populations and to reveal the patterns determining them, the results obtained by other workers in this field are treated only where they touch on this problem. The history of biological productivity has been exhaustively treated in a number of reviews (Ivlev, 1945, 1964; Zenkevich, 1947; Karzinkin, 1952;

Vodyanitskii, 1954; Vinberg, 1965). Topics relating to the classification of water bodies according to their productivity are also not
treated here (Bogorov, 1966, 1967 a, c; and others). The handbook
"Methods for Determination of Production of Aquatic Animals"
(Vinberg, 1968 a) can be regarded as the starting point of this work
(not chronologically, but rather with respect to its content).

During the preparation of the manuscript valuable advice and
remarks were given by G. G. Vinberg, V. A. Vodyanitskii, V. N. Greze,
T. S. Petipa, M. E. Vinogradov, L. M. Sushchenya, E. V. Pavlova and
Z. Z. Finenko, to whom I am sincerely grateful.

Chapter I

GENERAL CONCEPTS OF THE PRODUCTION
PROCESS IN BIOSYSTEMS

The material presented in this chapter serves as a brief introduction to the problem of biological productivity. This chapter
deals with theoretical concepts of biological productivity used in
this work. An attempt has been made to summarize the theoretical
conclusions of different workers in connection with general problems
of productivity: statement of the problem, definition of the different
terms, the relation between indexes used in working out production
characteristics and all work related to gathering factual data.

The different problems are treated as an internally coherent
system of concepts. Once established, definitions of the different
terms are strictly preserved in different situations. One of the major
sources of difficulty in some fields of biology, especially production
biology, is the arbitrary and ambiguous use of terms. Because of
this, mathematical treatment was the only possible way of overcom-
ing existing controversy and ambiguity. Some previously unex-
plored topics will be discussed on the basis of current views.

1. BASIC CONCEPTS AND PARAMETERS

As an object of study, a biosystem at any level of organization
presents a great number of scientific problems. The investigator
encounters the problem of productivity at the moment he sets out
to determine at what potential or actual rate a given biosystem
forms organic matter using assimilated compounds and energy.
The result can be expressed in units of mass or energy equivalents
or, with some reservation, in numbers of individuals.

Productivity of a biosystem or ecosystem refers to its capacity
to produce organic matter. Systems of the same type may have
different **levels of productivity** which can be judged from the magnitudes of **production indexes.** The basic production indexes are
production (P) and and **specific production** (C). Both indexes are

calculated per unit time (day, month, year). The **production** refers to all organic matter produced by a given system, regardless of whether it remains in the system or is partly or completely eliminated by the end of the test period. The production of a population consists of production from both growth and reproduction (total weight of eggs or larvae). These components can also be called production by a somatic and generative growth.

Specific production means the production in unit time per unit of biomass. With some reservation, the specific production can be considered synonymous with the ratio P/B.

One of the major tasks of production biology is to compare the productivity of biosystems. Naturally, only the same types of system can be compared (it is ridiculous to pose the question of whether a given population or a given community is more productive). The comparison can refer to the productivity of the whole systems within their natural boundaries, or be calculated according to a unit of space (area or volume) occupied by the system. Comparisons of the productivities of systems should be based on actual production indexes. Production is the most important of these indexes. It must be taken into account that the magnitude of production is a function of the specific production and biomass of the system. The production can be easily assessed from biomass dynamics and the nature of fluctuations in specific production of the system examined. Since biomass dynamics of the system must be determined in any such quantitative hydrobiological work (such material is quite abundant in the literature), the investigator of productivity must inevitably deal with the specific production. I believe that, at present, the study of the patterns determining the various specific productions of a system can provide a key to many problems of productivity. This is the theme of the present work, which essentially deals with the evaluation of the specific productions of different populations and the search for the patterns controlling their quantities.

I prefer not to use the term "P/B ratio," and favor using "specific production," especially when referring to diurnal periods. This unconventional approach requires some clarification. The term "P/B" was introduced (Zenkevich, 1931)* at a time when assessments of the annual production of large communities (plankton, benthos, fish) were only rough estimates. This ratio assists in clarifying discrepancies between biomass and production, and in establishing the direction of productivity fluctuations through a series of trophic

---

* Demoll (1927) used a similar expression (P/B), but in a different sense – namely, as the ratio between the production of animals (P) to the biomass of food organisms (B). This expression is similar to the F/B coefficient (Alm, 1924).

levels in marine communities. The designation of this index reflects the method of its calculation, first, annual production is evaluated and then divided by the biomass to obtain the annual P/B ratio. For long time intervals, however, many workers began to relate production variously to initial, average or minimal biomasses, causing confusion because results could not be compared. At present, it is recommended to calculate P/B for the shortest possible time intervals, and to relate production to the average biomass (Vinberg, 1968 a).

Detailed examination of the topic revealed that the diurnal magnitude of P/B in a population actually represents the weighted mean of the specific diurnal weight increments of all individuals of the population. This means that the magnitude of P/B of a population can be judged from data on individual growth rates and the age structure in the population. This implies that production indexes be determined in a different order. Earlier, calculations were made in the following sequence: biomass — calculation of production over a long time interval — calculation of the P/B ratio by dividing production by biomass. Now, however, the following order appears far more effective: biomass — calculation of diurnal P/B ratio according to data on individual growth rates and age structure — assessment of production from B and $C = P/B$. Because of mathematical difficulties in using the P/B expression, I have substituted the term "specific production" designated by C. It should be remembered that Vodyanitskii (1954) and Kamshilov (1958) introduced the term specific production for this index.

2. MATHEMATICAL EXPRESSION OF THE BASIC QUANTITIES AND THE RELATIONS BETWEEN THEM

Strictly productional studies, i. e., calculations of the different indexes on the basis of original biological data, involve purely mathematical operations; this makes the mathematical expression of terms and indexes important. Many works in the field of production biology are almost worthless because of errors made in the algorithm for the calculation of an index. Apart from these errors, some terms may lose their initial sense because of vague definition. In assessing the production of zooplanktonic populations, for example, Geinrikh (1956) calculated a quantity other than production. Ivanov (1955) proposed a method for calculating bacterial production that actually yields a different quantity. Many microbiologists have used this method without noticing this discrepancy.

For this reason, I decided to use certain "mathematical" terms for the main concepts of the production process, based on previously published works (Clarke et al., 1946; Ten and Zaika, 1967). The relations discussed here are widely used in the following chapters.

Basic Notations and Definitions

$t°$ – temperature in °C

t – time. This symbol designates a moment of time or a time period ($t = t_2 - t_1$) which is always clear from the context

τ – age of the individual

w – weight of the individual aged τ

w_0 – initial weight of the individual during the postembryonic period

w_∞ – weight limit, theoretically reached at $\tau \to \infty$ (used in relation to the metabolic equation of Bertalanffy)

w_m – actual weight limit of the individual

R – rate of food consumption by individual (or population)

A – rate of food assimilation

T – rate of expenditure for energic metabolism

τ_m – maximal life span of the individual

$\dfrac{dw}{dt}$ – rate of weight increase

q_w – specific rate of weight increase $\left(q_w = \dfrac{dw}{dt} \cdot \dfrac{1}{w} \right)$

q_l – specific rate of linear increase

K_1 – coefficient of utilization of food consumed for growth $\left(K_1 = \dfrac{dw}{dt} \cdot \dfrac{1}{R} \right)$

K_2 – coefficient of utilization of food assimilated for growth $\left(K_2 = \dfrac{dw}{dt} \cdot \dfrac{1}{A} \right)$

N – number of individuals

n – distribution of the number of individuals by age $\left(n = \dfrac{dN}{d\tau} \right)$

E – rate of elimination of individuals

B – biomass of the population (B_t is the biomass at moment t)

B'_e – biomass of eliminated individuals

B''_e – biomass of organic matter eliminated during life (molts, gland secretions, etc., but not in the form of genital products)

B_e – total biomass eliminated $\left(B_e = B'_e + B''_e \right)$

P – production (production rate)

P_t - production during the period $t = t_2 - t_1$

P_1 - production due to reproduction (total weight of eggs or larvae)

P_2 - production due to individual growth

C - specific production (production rate)

The above list includes only the symbols most commonly used in this work and which appear in many of its chapters. Other designations used in a specific sense in different chapters are explained in the text.

The following mathematical expressions of production and specific production will be used as definitions of these concepts:

$$P_t = B_t - B_0 + B_e,$$

$$\tag{1}$$

$$C = \frac{P}{B}.$$

$$\tag{2}$$

As will be seen, these indexes can be expressed in other ways, one of which can serve as a convenient definition of the given quantity.

Most of the indexes listed above reflect the rate of different processes at time t, although their names, following the established practice, do not always reflect this fact. It would be more correct to use symbols such as $R_{(w,t)}$ for the rate of food consumption by an individual of weight w at time t or $P_{(w_m,t)}$ for the production rate of individuals ranging in weight from w_0 to w_m at time t; this approach allows analysis of part of the population, e.g., the notation $P_{(w,t)}$ indicates that only individuals weighing from w_0 to w are concerned.

The rates readily yield the respective quantities for the period $t = t_2 - t_1$. Thus

$$P_t = \int_{t_1}^{t_2} P(t) \, dt.$$

$$\tag{3}$$

As noted above, it is not always necessary to use the term "production rate" for P; in equation (3), however, P represents precisely the production rate, whereas P_t is the production during time $t = t_2 - t_1$.

Equation (1) is taken as the mathematical definition of the production, which corresponds with the interpretation of this term by Boysen-Jensen (1919). However, other methods for its expression also exist. Since it is not always evident how these methods pertain to the definition of production, it would be desirable to compare

the different formulas. Such an analysis can also reveal the relation between production and current ecological views on population
growth.

Let us visualize the population as a system whose input is the
rate of biomass formation, characteristic for the given system, and
whose outputs comprise the dynamics of fluctuations in
existing biomass and the biomass elimination rate (Figure 1).

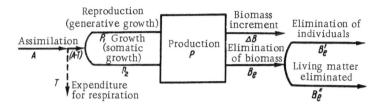

FIGURE 1. Diagram illustrating conversions of matter and energy
assimilated by a population

1. According to the definition, production is assessed from the
output of the system. With respect to the time interval $\Delta t = t_2 - t_1$,
equation (1) can be expressed in the form

$$P = \Delta B + B_e, \qquad (4)$$

where $\Delta B = B_2 - B_1$. As $\Delta t \to 0$ we obtain

$$\frac{dP}{dt} = \frac{dB}{dt} + \frac{dB_e}{dt}. \qquad (5)$$

If the existing biomass does not change with time, it follows that

$$\left.\begin{array}{l} \Delta B = 0, \\ P = B_e. \end{array}\right\} \qquad (6)$$

When assessing production from the output of the system, it should
be remembered that the biomass increment (ΔB) is calculated as the
change in total weight of live organisms, whereas the eliminated
biomass (B_e) consists of two components: the individuals eliminated
(B_e') and the organic matter formed in the given system and eliminated in life processes (B_e''), i.e.

$$B_e = B_e' + B_e''. \qquad (7)$$

Thus, elimination of biomass does not necessarily mean death of organisms. Production not only includes the "live biomass" increment (ΔB), but also individuals (B_e') eliminated as a result of predation or natural mortality, as well as molts, secretions, etc. ($B_e^{\cdot}$); for consistency, these designations are used in defining production according to long-established practice (Thienemann, 1931; Borutskii, 1939 a, b).

2. In recent years, Soviet hydrobiologists have determined production of a population mainly from the input of the system (see Figure 1):

$$P = P_1 + P_2. \tag{8}$$

For calculating production on the basis of equation (8), one considers the average abundances of different age groups for the period Δt, and so the changes in initial abundances and biomass during the indicated period will be reflected in the average values. As a result, the calculations of the production according to "input" (equation 8) and according to "output" (equation 1) give the same result.

3. Approaching the problem of production from a physiological viewpoint, it is easily seen that, owing to the law of conservation of energy, production equals the difference between the assimilated energy (A) and metabolic expenditure (T); this yields yet another expression of the production in terms of the input of the system (see Figure 1):

$$P = A - T. \tag{9}$$

4. Rewriting equation (2) as

$$P = CB \tag{10}$$

we again have a correct expression for production which will be extensively used in our work.

5. Finally, production can be calculated from input of the system only from dynamics of population without taking into account individual growth. Such an approach may be applied to any organism, but it is unavoidable for microorganisms. In this case, all individuals are regarded as having some average weight $\bar{w}$, so as to deal with population sizes without recourse to the age structure (all individuals are regarded as interchangeable and equivalent). Through average weight, population size readily yields biomass. In this case, the "input" of the system would be the reproduction rate of the population (the increase in biomass results from an increase in population).

The condition described above is equivalent to absence of individual growth ($P_2 = 0$); thus equation (8) reduces to the form

$$P = P_1.$$

Works dealing with the population dynamics usually give an exponential model of population growth (here biomass is used in place of population size). According to this model the rate of population increase $\left(\frac{dB}{dt}\right)$ is proportional to the existing biomass:

$$\frac{dB}{dt} = rB, \tag{11}$$

$$r = b - m,$$

where r is the coefficient of population increase ("intrinsic rate of increase" used by British and American authors), b the reproduction coefficient and m the elimination coefficient.

Where elimination does not occur, it follows from equations (5) (5) and (11) that

$$\frac{dP}{dt} = \frac{dB}{dt} = bB. \tag{12}$$

Thus, if the increase in biomass results entirely from reproduction and elimination does not occur, the production rate equals the biomass increment.

Using the definition of specific production in equation (2), we obtain from equation (12) that

$$C = b, \tag{13}$$

i.e., specific production equals the reproduction coefficient. To demonstrate that equation (13) applies not only where there is no elimination, equation (11) can be written as follows:

$$\frac{dB}{dt} = bB - mB. \tag{14}$$

Here mB is the rate of biomass elimination; the same index appears in equation (5) as $\frac{dB_e}{dt}$:

$$\frac{dB_e}{dt} = mB. \tag{15}$$

Comparing equations (5), (14) and (15) we obtain

$$\frac{dP}{dt} = bB.$$

This is identical with equation (12) where $C = b$.   Thus, expression (13) applies under any conditions.

3. THE CONCEPT OF PRODUCTION AS APPLIED
TO DIFFERENT TYPES OF BIOSYSTEMS

"... Soviet authors commonly apply the term 'productivity' to a characteristic feature of a given population, community or water body which is expressed in a certain — small or large — magnitude of production ..." (Vinberg, 1968 a).   From this citation it follows that production refers to such biosystems as population and community; moreover, this term applies as well to whole water bodies, i. e., to large ecosystems.

In terms of individuals, the problems of productivity relate to physiology of growth and energy balance.   Nevertheless it is worthwhile to define concepts of production biology on the individual level.   This approach can shed light on the change in the actual expression of production terms in transition from individuals to increasingly complex biosystems, including ecosystems of different ranks.

INDIVIDUAL PRODUCTION

In recent years, production studies have drawn the fields of ecology and physiology closer.   "Productionists" can use a rich store of physiological data and laws, whereas physiologists have an interesting and important task which stimulates the development of ecological-physiological research.   Because of the mutual influence of these two fields, production indexes have recently been calculated for short periods (24 hours) which facilitates comparison with physiological rates.   At the same time, physiologists are becoming interested in the course of energy processes not only for given periods of individual life, but also throughout the course of ontogenesis, which allows more adequate evaluation of the role of production of the individual as an element of the population.

It was found that certain physiological relations require clarification.   The need for detail applies in particular to the widely used balance equations based on the laws of conservation of matter and energy (Vinberg, 1956, 1962):

$$R = \frac{dw}{dt} + T + D,$$

$$A = \frac{dw}{dt} + T,$$

(16)

where D is the unassimilated part of the ration.

If the quantities occurring in equation (16) are measured in short-term experiments, the results agree satisfactorily with the balance equation. However, if the same balance elements refer to the entire life span of the animal, there will be no equality if P_1 and $B_e^{''}$ are disregarded. This is stressed in the handbook (Vinberg, 1968 a), as well as in the paper of Khmeleva (1968).

It follows that equation (16) is more accurately written

$$A = \frac{dw}{dt} + P_1 + B_e^{''} + T.$$

(17)

The term "increment" has a dual sense: it is used to designate the rate of weight increase $\left(\frac{dw}{dt}\right)$ as well as the quantity $A - T$. In the first case, the quantities P_1 and $B_e^{''}$ are treated as "losses," whereas in the second case the growth rate will be $\frac{dw}{dt} + P_1 + B_e^{''}$. Two processes are present which should be designated with different terms. Unlike the conventional growth rate measured as the value of $\frac{dw}{dt}$ it is desirable to introduce the term "individual production," measured by the term $A - T$ or $\frac{dw}{dt} + P_1 + B_e^{''}$, which completely corresponds to the production on the population level. We note that in the case of populations, the increment of existing biomass is also not always equal to production (Petrusevich, 1967).

In production studies in general, it is necessary to proceed from individual production as defined above. However, if the study of all three components of the production rate, namely $\frac{dw}{dt}$, P_1, and $B_e^{''}$, is difficult for some reason, then the individual production rate can be assessed approximately from $\frac{dw}{dt}$ or $\frac{dw}{dt} + P_1$. This deliberate simplification rests on the assumption that $\frac{dw}{dt} + P_1 \gg B_e^{''}$ or that $\frac{dw}{dt} \gg P_1 + B_e^{''}$; accordingly, $A - T \approx \frac{dw}{dt} + P_1$ or $A - T \approx \frac{dw}{dt}$. Errors arising from such a simplification in the calculation of production can only be determined by corresponding study of representatives of different groups. Presumably, the values of P_1 and $B_e^{''}$ cannot always be treated as negligibly small. In particular, for estimating the production of microorganisms it is necessary to remember that vital elimination of organic matter may amount to a fairly large proportion of total production (Khailov, 1969).

PRODUCTION OF POPULATIONS

In biosystems of this level the basic concepts retain the same definitions at the individual level, but are estimated on the basis of somewhat different elements. For the individual, egglaying involves a loss of biomass from the system, whereas on the population level the progeny remain within the system and enlarge the population. The death of an individual within a population does not mean destruction of the system, but merely loss of biomass. Most production studies refer to whole populations. All that has been said on production in sections 1 and 2 refers particularly to populations.

One theoretical aspect deserves attention. Strictly speaking, production of a population cannot be treated as the total of all individual productions in the presence of cannibalism, when young mammals suckle or in other similar situations. This also applies where the production of a community composed of organisms of two trophic levels is evaluated. As will be explained below, production of such systems is not equal to the total of the productions of all the populations (Vinberg, 1936).

PRODUCTION OF SUPRAPOPULATIONAL SYSTEMS

Ecologists distinguish between different types of suprapopulational systems, ranging from two interacting populations to major ecosystems as lakes or seas. If a system composed of a single trophic level is studied, then its production is equal to the sum total of the productions of the populations constituting the system.

The analysis of the production is more difficult for more complex systems. There are many general discussions published of productivity of various communities and ecosystems based on biomass estimates, but very few concrete calculations of the production itself. Some aspects of the productivity of large systems with reference to existing classifications of communities and ecosystems have been discussed (Zaika, 1967 a). Here some possible approaches to the production of suprapopulational systems will be considered.

Vinberg (1936) indicated that the total production cannot be obtained from the sums of the productions of the populations which constitute a community composed of two or more trophic levels. Shushkina (1966) assumed that lake zooplankton consists of two trophic levels and calculated its production according to the equation

$$P = P' + P'' - R'',\qquad(18)$$

where P is the total production of the system, P' the production of the lower trophic level, and P'' and R'' the production and ration of the higher trophic level. Shushkina noted that the equation can be used for determining production of the zooplankton **which may serve as food for fish.**

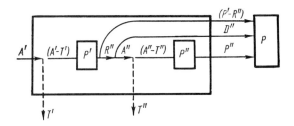

FIGURE 2. Diagram illustrating conversions of matter and energy in a system composed of organisms of two trophic levels (explanations in the text)

Indeed, equation (18) requires this qualification because it does not quite correspond to the concept of "production." This reservation is explained in Figure 2 which shows the elements of a production process in a system composed of two trophic levels. Obviously, equation (9) applies to each trophic level as well as to the system as a whole. In the latter case

$$P = A' - T' - T'',\qquad(19)$$

because A' is the amount of matter entering the system, whereas the losses are determined as a sum $(T' + T'')$. Since $P' = A' - T'$ and $P'' = A'' - T''$, equation (19) can be transformed to

$$P = (P' + T') - T' - (A'' - P''),$$
$$P = P' + P'' - A''.\qquad(20)$$

A comparison of equations (20) and (18) shows that the "internal losses" of the population are expressed by A'' rather than R''.

In other words, Shushkina (1966) ignored the fact that D'' (the unassimilated part of R'') is also a component of the production of the given system.

The same result can be obtained in another way. Using Figure 2, an estimate of production of the system "at the output" is

$$P = (P' - R'') + D'' + P''. \tag{21}$$

This readily leads to equation (20).

Certain reservations have been expressed as to the validity of the above interpretation for a system of two trophic levels. Certain additional explanations are therefore necessary. The main reservation is that the production of the system, as defined in equation (20), includes the feces of the second (higher) trophic level. This is supposedly untrue because unassimilated food is not included in production of a population. Moreover, it is claimed that our interpretation is a departure from estimation of production in terms of living organic matter.

From our viewpoint, all the reservations can be rejected for the following reasons.

1. In any case, the system receives from without an amount of matter equal to the assimilation. Naturally, in a population the unassimilated food does not constitute part of A and should not be considered in production estimates. In a system of two trophic levels, the unassimilated part of the food of the first level must be similarly disregarded. The food of the second trophic level is by definition derived from within the system and not from without, i. e., from the production of the first trophic level. The only loss relates to that part of the ration of the second level which is used for metabolism. The rest of the ration (P'' and D'') is a component of the total production of the system.

2. The definition of production does not refer to "live" matter only. By definition, organic matter created in the system is treated as production regardless of its fate. Estimates of the production of populations include exuviae and other formations not constituting "live" matter (Thienemann, 1931; Borutskii, 1939 a, b).

3. Since the quantity of production is not related to its further utilization, the calculations of the productions available for **fish, man or bacteria** always yield different values for the same system if based on the general determination of production. Such calculations are necessary because they serve important specific tasks; however, they indicate only **part of the production of the system.**

Thus, the successive application of the concept of "production" to a system of two trophic levels leads to equation (19). In the presence of n levels the production of the system can be expressed in a similar way, taking into account the losses on all levels.

Vinberg (1936) interprets productivity of a water body in the light of the relation between two opposite trends — primary production and destruction (the latter can be calculated as the total consumption of energy by the population of the water body). In discussing possible situations, Vinberg writes: "... It is obvious that the balance of organic matter can be positive, zero or negative. A zero balance resulting from an annual cycle is only possible in a water body with a fully reversible metabolism which is a theoretical abstraction. Real water bodies, however, may more or less approximate this extreme condition ..."

Thus, Vinberg regards primary production under a zero balance as the total of expenditures of all organisms for respiration. This can be expressed by the equation

$$A' = T' + T'' + \cdots + T^n, \tag{22}$$

where A' is the assimilation of the primary producers (so-called gross primary production), T', T'', ..., T^n the losses of the consecutive trophic levels from 1 to n. MacFadyen (1964) and Odum (1959) treat this problem similarly.

As noted above, equation (19) assumes the following form for a system of n trophic levels

$$P = A' - T' - T'' - \cdots - T^n. \tag{23}$$

According to equations (23) and (22), the production of the system equals zero in the ideal situation of a zero balance of organic matter. Vinberg (1936) analyzes conditions under which a water body can have a positive or negative balance of organic matter. The production is more difficult to assess in such cases. In particular, if the production of a water body with a negative balance is determined from equation (23), the value of A' cannot be defined solely as the gross primary production since a negative balance can also arise when allochthonous organic compounds reach consumers without passing through the producers.

The productivity of communities and ecosystems is analyzed using various indexes characterizing different aspects of processes taking place within the complex systems. These indexes do not always correspond to production indexes used in population studies. The characteristics of complex communities are evaluated as a pyramid of productions and specific productions. Another index is the ratio between the primary production and total losses at all levels (Odum, 1959). In analyzing successive changes within a community, Margalef (1960, 1961, 1963 a) and others use as an index of productivity the ratio between primary production and biomass of the whole community.

Thus, the concept of "production" assumes different concrete expressions in different types of systems, but its fundamental definition remains unchanged.

Chapter II

ANIMAL GROWTH— THE BASIS OF
PRODUCTION OF POPULATIONS

1. RELATIONSHIP BETWEEN GROWTH AND PRODUCTIVITY OF ANIMALS

The production process in a population is represented in the biomass increment which includes the individual weight increase and increased number of individuals by reproduction. Several methods for calculating production of a population were indicated in the preceding chapter. These methods differ with respect to the basis of the calculation of production which was in one case the ratio between total assimilation and metabolic losses, in another case the population dynamics or individual growth, and so on.

The difference between approaches based on population dynamics and on individual weight increase is particularly instructive. In the former case, growth is disregarded and production is interpreted as an increase in abundance of identical organisms; the second approach is based on growth and considers reproduction as a continuation of individual growth. Although the approaches are essentially similar, production is now principally determined on the basis of weight increase. This is because studies of the weight increase and age structure of a population require less effort and time and usually yield more reliable data for production calculations, compared with the laborious preparation of "life tables" and their application in determining the rate of population growth. Soviet scientists have worked out exact procedures for determining production on the basis of growth data (see Chapter III). This approach also permits the application of advances in work on animal growth.

"... Production in the population of a species involves the individual increments of organisms constituting the population, including the increment of sexual products and other organic formations that have been detached from the organisms during the test period. Therefore, for determination of production it is necessary to have quantitative data on growth, the length of different development stages, fecundity, as well as information on their relation to

19

environmental conditions. This is to be guided by general concepts on types of growth" (Vinberg, 1968 a).

A comparison of these considerations with those described in Chapter I on the relation between individual growth and production must include "individual increments" which here represent the individual production. Part of this production is the growth increase proper, as reflected in individual growth curves. Existing concepts on the main types of growth were worked out on the basis of an analysis of growth curves, and are practically unrelated to individual production. After growth ceases, animals begin intensive production of sexual products; because of this the overall pattern of the "production curve" may differ markedly from the growth increase curve. Figure 3 shows growth and production curves of D a p h n i a . The production curve should actually be even higher because it does not reflect other categories of metabolic losses of organic matter.

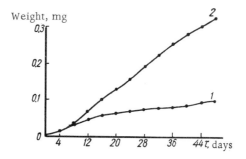

FIGURE 3. Growth and production of female D a p h n i a p u l e x :

1) weight increase; 2) individual production (total of body weight and weight of all eggs laid at the given time). After Pechen' and Kuznetsova (1956).

The growth curve makes it possible to determine absolute and specific daily increments of individuals of different ages (weights); these data serve for calculating the production of the population. The mathematical description of the growth curves facilitates both determination of specific growth rates and the theoretical analysis of the effect of different factors on production of the population. This will be demonstrated for some aspects of growth as outlined by Vinberg (1966, 1968 a, b).

The specific production of a population is based on the individual specific growth rates. The average specific rate of weight increase q_w during the period $t_2 - t_1$ can be calculated according to the formula

$$q_w = \frac{\ln w_2 - \ln w_1}{t_2 - t_1},$$ (24)

where w_1 and w_2 are the weights at times t_1 and t_2, respectively. Equation (24) applies to any type of growth, but it remains empirical because it is impossible to forecast the behavior of q_w as a function of the weight attained. A mathematical presentation of growth can solve this problem.

For exponential growth we have

$$\frac{dw}{dt} = q_w w$$ (25)

and

$$q_w = \frac{dw}{dt} \cdot \frac{1}{w} = \text{const},$$ (26)

i. e., if we know that growth is exponential repeated calculation of q_w is no longer necessary, since the specific growth rate in this case remains constant during growth.

In a parabolic type of growth

$$\frac{dw}{dt} = aw^b,$$ (27)

where a and b are coefficients; this leads to

$$q_w = aw^{1-b}.$$ (28)

It can be seen that q_w decreases proportionally to w^{1-b}. If the growth curve is described mathematically, and a and b are known, q_w is easily calculated for any value of w.

The possibility of determining production through a profound analysis of growth is very attractive and deserves wide application. However, this only estimates production due to growth, i. e., P_2, and does not refer to P_1 or B_e'. While B_e' is often negligible, P_1 should be determined separately and then added to the production resulting from growth (equation (8)). Greze and Baldina (1964) proposed plotting individual growth curves similar to that in Figure 3 in production assessments. Although acceptable, this procedure hinders using mathematical relations relating to growth. It is especially necessary to note that relating equation (24) to the production curve yields the quantity $\frac{\Delta P}{\Delta t} \cdot \frac{1}{P}$ which is often much smaller than $\frac{\Delta P}{\Delta t} \times \frac{1}{w}$, and therefore cannot be used for calculating the specific production of a population. Therefore P_2 and P_1 are determined separately in assessing production by Greze's method (Zaika, 1969 a).

Indeed, if the value of $q_w = \dfrac{\ln w_2 - \ln w_1}{t_2 - t_1}$ is calculated according to the production curve in Figure 3, we obtain $q \approx 0.04$ for daphnias aged 30 days. This index shows the ratio between the production rate and the **production** of daphnias for 30 days, amounting to about 0.208 mg. However, this value does not interest us in calculating the specific production; instead, we strive to find the ratio between the production rate and existing biomass, i. e., the **weight** of the daphnias which equals 0.07 mg at this age. Accordingly, the specific production per 24 hours is 0.11 rather than 0.04.

Discussing the relation between individual growth and production, we concluded that the production of a population should be assessed by determining P_2 on the basis of the growth curve, and then adding it to the value of P_1 which is determined separately ($B_e^{'}$ must also be assessed whenever possible). The plotting of an individual production curve can yield much valuable information in production research. Because of this, the analyses of individual growth and production curves are interrelated, but at the same time are independent tasks requiring profound study.

This chapter presents results of our studies on animal growth as an extension of the theoretical research of Vinberg (1966, 1968a, b), who has made a number of important generalizations and has attracted the attention of biologists to this old problem.

As Vinberg (1966) showed, the best mathematical expression of the process of individual weight increase is the weight increase equation which was developed and analyzed by a number of authors, notably Pütter (1920), Bertalanffy (1938) and Taylor (1960). The equation is usually linked with the name of Bertalanffy because of his major contribution to this problem. In its most general form the Bertalanffy equation can be written

$$\frac{dw}{dt} = a_1 w^{b_1} - a_2 w^{b_2}, \tag{29}$$

where w is the individual weight, $\dfrac{dw}{dt}$ the rate of weight increase (increment), and a_1, a_2, b_1 and b_2 are coefficients. According to Bertalanffy, the expression $a_1 w^{b_1}$ on the right of the equation measures "anabolism" which is proportional to the body surface, hence $b_1 = \dfrac{2}{3}$. The second term, $a_2 w^{b_2}$, reflects "catabolism," which is proportional to body weight, $b_2 = 1$. Accordingly, the Bertalanffy equation is usually used in the following form:

$$\frac{dw}{dt} = a_1 w^{2/3} - a_2 w,$$ (30)

although, for a number of reasons, the condition $b_1 = \frac{2}{3}$ is often rejected.

The main shortcoming of equations (29) and (30) is the fact that their parameters lack clear biological definition, making it impossible to study the coefficients (a_1, a_2, b_1, b_2) independently. Such concepts as "anabolism," "catabolism" and "surface" (active surface with respect to anabolism) are quite vague. There were attempts to make the Bertalanffy parameters more concrete or replace the equation by others based on different considerations. Vinberg correctly pointed out, however, that the other mathematical expressions for growth proposed so far have no advantages over the Bertalanffy equation and often have serious disadvantages.

The **theoretical foundations** of the Bertalanffy equation clearly require further analysis, but there can be no doubt as to its **applicability to growth** of a large number of animals. This equation is sufficiently flexible; in the form (30) it can be applied to S-type growth (attenuated growth with a relatively clear-cut limit); if $a_2 w = 0$ it describes parabolic growth, and if $b_1 = 1$ it represents exponential growth.

It should be emphasized that the application of the Bertalanffy equation to parabolic and exponential growths clearly demonstrates all the faults in the initial interpretation of the parameters. Indeed, for describing parabolic growth it is necessary that $a_2 w = 0$. This condition means that the catabolism equals zero, which is a biological absurdity. At the same time, it is also necessary to abandon the condition $b_1 = \frac{2}{3}$; otherwise it becomes impossible to characterize many cases of parabolic growth or exponential growth. This largely discredits the reasoning that anabolism is proportional to the body surface, i. e., weight to a power of $\frac{2}{3}$.

On the basis of these remarks, I shall attempt to explain my position with respect to the Bertalanffy growth theory and the character of my research on animal growth. Since the Bertalanffy equation characterizes growth satisfactorily, it has been extensively used in my work, especially in studies on the growth of Larvacea [Oikopleura], sagittas, mollusks, etc. Our mathematical models were easily derived from various implications of the Bertalanffy

growth theory (specific growth rate as a function of weight, etc.).
All these applications of the Bertalanffy equation have a special
significance, not related to the interpretation of the equation parameters and discussed in the chapters in which they are used to solve
actual problems (Chapters V and VI).

The results of studies of the quantitative relationship between
specific growth rates of newborn animals and their weight is discussed in this chapter, as well as the new, more concrete significance proposed for the parameters of the Bertalanffy equation.
Although relating to the problem of growth, all these topics have
important implications for productivity.

Although warm-blooded animals do not immediately interest us,
data on their growth are central in the analysis of the relation
between the specific growth rate of newborn animals and weight.
This is because poikilothermic animals are less suitable for testing
a distinct quantitative relationship between these quantities. First,
they are highly heterogeneous phylogenetically; second, data must
be obtained at equal or similar external temperatures; third, exact
measurements of short-term weight changes of newborn invertebrates are difficult to make because of their small size, and therefore are rare. For this reason only a few data on the growth of
cold-blooded animals are used.

2. CORRELATION BETWEEN MAXIMAL SPECIFIC
GROWTH RATES OF ANIMALS

At present, a considerable amount of data is available on growth
in many animals. Repeated comparisons have been made of growth
rates in representatives of various animal groups. Since animals
radically differ in weight and duration of the growth period, it is
very important to select a suitable comparative index to characterize the growth rate. One of the best indexes for this purpose is
the specific growth rate q_w, calculated according to equation (24).
This index can be used to compare the growth of animals. It is
known that the specific growth rate usually decreases during growth.

A comparison of the specific growth rates of newborn animals may allow finding the maximal specific growth rates q_m for the post-embryonic period. Examination from this viewpoint of the vast data on growth of warm-blooded animals revealed a distinct quantitative relationship between newborn weight and q_m (Zaika, 1970). The animals studied belong to 4 groups, each having characteristic values of constants for the general equation

$$q_m = pw^n, \qquad\qquad (31)$$

where p and n are coefficients:

Mammals (other than primates) . . . . $q_m = 0.142\ w^{-0.102}$
Primates . . . . $q_m = 0.021\ w^{-0.213}$
Altricial birds . . . . $q_m = 0.435\ w^{-0.139}$
Nidifugous birds . . . . $q_m = 0.274\ w^{-0.250}$

Analysis of these data leads to the conclusion that the differences between these groups of warm-blooded animals are due to the fact that the organisms are born at different stages of development; the more developed the newborn animal is the lower is its q_m at an equal w. It must follow that the "true potentialities" of growth are most evident in animals born less mature, i.e., in altricial birds and mammals (except primates). It must be admitted that these two groups were best studied with respect to q_m. It was assumed that if the curves relating q_m and w for these two groups according to equation (31) were continued to the left, i.e., toward lower weights, they would lie within the region of points representing the specific growth rates of cold-blooded organisms, notably aquatic invertebrates, since the growth potentialities of all organisms must be rather close. The data on invertebrates and microorganisms confirm this assumption (Zaika and Makarova, 1971) (Figure 4). Without discussing topics related only to the growth problem, it is pertinent to note that the relationship established allows an approximate estimate of the upper limit of specific production of a population according to the size and weight of its individual members. Indeed, a population of any species will have a maximal specific production if it consists entirely of newborn individuals, since each individual will have the maximal possible specific growth rate.

Most populations have a rather complex age and size structure,
which has to be taken into account when evaluating specific production. When determining the specific growth rate and specific production, it should be remembered that animals cannot have q_w higher

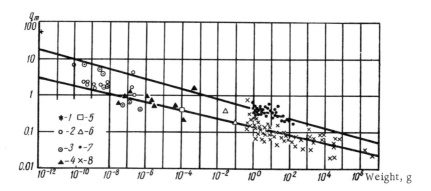

FIGURE 4. Correlation between maximal specific growth rates (q_m) of
organisms of different weight (w):

1) bacteria; 2) unicellular algae; 3) infusoria; 4) arthropods; 5) worms;
6) amphibians; 7) altricial birds; 8) mammals. The lines refer to altricial birds and mammals (after Zaika and Makarova, 1971 b).

than those determined for animals of a given weight according to
equation (31) for constants typical, say, of altricial birds, i. e.,
$q_m = 0.435 \, w^{-0.139}$.

3. BIOLOGICAL SIGNIFICANCE OF THE PARAMETERS
OF THE BERTALANFFY GROWTH EQUATION

An explanation of why the Bertalanffy equation should be regarded
as a convenient empirical expression of growth, rather than a
theoretically justified formula based on concrete and proven
theoretical considerations, was given in the introduction to this
chapter.

In an attempt to improve the significance of the terms in the equation, Paloheimo and Dickie (1965) started from the energy balance equation of Vinberg (1956)

$$\frac{dw}{dt} = A - T, \tag{32}$$

according to which the increment equals the difference between assimilated food and metabolic losses. It is known that these losses are related parabolically to individual weight:

$$T = a_2 w^{b_2} \tag{33}$$

(here and below a and b with different indexes are constants). Hence

$$\frac{dw}{dt} = A - a_2 w^{b_2}. \tag{32'}$$

In order to find the functional relation between the increment and the weight reached, it is sufficient in equation (32') to express A as a certain function of w, which Paloheimo and Dickie were unable to do.

Independently of these authors, Vinberg (1966, 1968 a, b) also used equations (32) and (33) to derive a relationship between growth and metabolism: if the ratio v between $\frac{dw}{dt}$ and T remains constant during growth, an equation of parabolic growth will be obtained:

$$\frac{dw}{dt} = vT = va_2 w^{b_2}. \tag{34}$$

To apply this approach to cases of attenuating growth, Vinberg assumed that the value of v may decrease during ontogenesis, and he selected a function $v = f(w)$ which leads to the S-shaped curve of Bertalanffy.

Recent work (Inoue, 1964; Sushchenya and Khmeleva, 1967; Abolmasova, 1969) showed that for a large number of crustacean species the ration R is parabolically related to body weight:

$$R = a_3 w^{b_1}. \tag{35}$$

I believe that this constitutes an advance toward a clearer biological meaning of the parameters of the growth equation (Zaika and Makarova, 1971 a). By assuming that the assimilability of food remains at an average level throughout the growth we obtain

$$A = Ru = ua_3 w^{b_1}, \tag{36}$$

where u is the assimilability coefficient $(u < 1)$. By substituting a_1 for ua_3 we obtain from equations (32) and (36)

$$\frac{dw}{dt} = a_1 w^{b_1} - a_2 w^{b_2}. \tag{37}$$

The equation obtained resembles (29), but differs in the interpretation of the parameters and in the exponents of w. Equation (37) is based on the following assumptions: 1) balance equality (equation (32)), 2) equation (33) linking metabolic losses with individual weight, 3) equation (35) linking the ration and individual weight, 4) the assumption that the assimilability coefficient remains constant during growth. The balance equality needs no further justification; the general form of equations (33) and (35) was apparently established rather convincingly; the values of the coefficients for w, as well as the relationship between food assimilation and body weight, can be determined experimentally. Equation (37) is more flexible than (30), and therefore covers a greater number of growth types.

If $a_2 w^{b_2}$ indicates only the metabolic losses, then equation (37) measures the **individual production** rather than growth. I believe that the equation can also be applied to growth. This necessitates the further assumption that metabolic losses, as well as other losses of organic matter during life (in reproduction, molts, etc.), can also be expressed as $a_2 w^{b_2}$ (although with different values of a_2 and b_2).

The quantity w_∞, which will be referred to as the definitive weight in the following analysis of equation (37), has a different significance when one considers the production curve rather than individual growth.

Since we are interested in describing the process of individual mass increase, i. e., the cases in which $\frac{dw}{dt} \geq 0$, the following discussion will be confined to an analysis of the theoretical situations in which $a_1 w^{b_1} \gg a_2 w^{b_2}$.

It was established empirically that the values of A and T increase with the weight of the animal. Representing the functions of equations (33) and (36) in a single graph with logarithmic coordinates will yield straight lines (Figure 5); we describe three cases.

1) $b_1 = b_2$. This condition has physical meaning only if $a_1 > a_2$. Lines A and T are parallel; according to equation (34)

$$\frac{dw}{dt} = (a_1 - a_2) w^{b_1}. \tag{38}$$

This is an equation of parabolic growth. At $b_1 = 1$ growth becomes exponential.

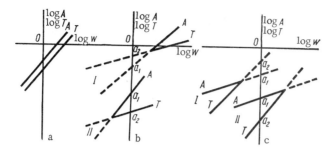

FIGURE 5. Possible relationships between assimilation (A) and losses (T) during growth:

a) $b_1 = b_2$; b) $b_1 > b_2$; c) $b_1 < b_2$; I) $a < a_2$; II) $a_1 > a_2$ (logarithmic coordinates). Explanation in text (see equation (40)).

2. $b_1 > b_2$. The lines intersect at a value of w which can be found from the relationship

$$w = \left(\frac{a_1}{a_2}\right)^{\frac{1}{b_2 - b_1}},$$
(39)

since $A = T$ and $\frac{dw}{dt} = 0$. Only the area to the right of this point has real meaning. Therefore, the initial weight w_0 of the individual cannot be less than w in equation (39); for example, if $a_1 < a_2, w_0$ must be greater than unity (see Figure 5). As in the preceding case, growth is unlimited here. The ratio v between increment and losses can be written as

$$v = \frac{a_1}{a_2} w^{b_1 - b_2} - 1,$$
(40)

i.e., this ratio is not necessarily constant in this type of growth. The same conclusion was reached earlier by other means (Zaika and Makarova, 1969).

3. $b_1 < b_2$. The area to the left of the intersection point given by equation (39) has actual meaning. In this case we have S-type growth and the definitive weight w_∞ can be found from equation (39). It is readily shown that $w_\infty < 1$ if $a_1 < a_2$ and that $w_\infty > 1$ if $a_1 > a_2$ (see Figure 5). The inflection point of the growth curve can be determined according to the equation

$$w_n = \left(\frac{a_1 b_1}{a_2 b_2}\right)^{\frac{1}{b_2 - b_1}},$$
(41)

and by equation (39)

$$w_n = w_\infty \left(\frac{b_1}{b_2}\right)^{\frac{1}{b_2-b_1}} . \qquad (41')$$

The form of the curves, obtained in all three cases, was tested by us in numerical examples. Equation (37) is a binomial differential which can be expressed in the following form

$$w^{-b_1}(a_1 - a_2 w^{b_2-b_1})^{-1} dw = dt. \qquad (37')$$

The integration of equation (37') is difficult, but quite possible for any concrete values of the coefficients; this allows the plotting of curves $w = f(t)$.

Let us point out once again that if the values of coefficients a_2 and b_2 in equation (37) are calculated on the basis of the oxygen up-take, especially in less active animals, the solution of this equation will not yield a curve of weight increase. According to the balance equation (32), equation (37) yields a growth curve only on the con-dition that the losses $a_2 w^{b_2}$ include not only the general metabolic losses (including active metabolism) but also all the other losses of organic matter (for reproduction, molts and secretory activity).

Thus, the substitution of "anabolism" and "catabolism" in the Bertalanffy growth equation by assimilation and losses expressed as a function of body weight yields a more flexible equation in which all the parameters can be experimentally checked; this equation not only characterizes production but also growth of the individual. The proposed methods for depicting growth and interpreting the parameters of the equation contsitute a natural development of the concept of growth based on new data on nutrition patterns. The main results of our analysis of growth do not essentially contradict the achievements of other growth researchers.

I presented the above considerations on the possibility of a new interpretation of the parameters of the Bertalanffy growth equation before a symposium in Minsk (June, 1969). During the discussion, G. G. Vinberg noted that if $b_1 > b_2$ the specific rate increases during growth to a maximum at

$$w = \left(\frac{a_1}{a_2} \cdot \frac{b_1 - 1}{b_2 - 1}\right)^{\frac{1}{b_2-b_1}} .$$

The question is whether such a condition is real, since an increase in the specific growth rate during growth is an extraordinary phenomenon.

Let us examine theoretically what the position is of maximum specific growth rate with respect to the critical value of w which limits the real area in the case $b_1 > b_2$ (remembering that there is real significance in the section where $a_1 w^{b_1} \geqslant a_2 w^{b_2}$).

If $b_1 > b_2 > 1$, the maximum specific growth rate will lie to the left of the critical point given by equation (39). In this case, therefore, the specific growth rate in the real area can **only decrease** with an increase of w. However, the condition $b_1 > b_2 > 1$ is biologically improbable. For $1 > b_1 > b_2$, the specific growth rate rises to a peak and then decreases. Thus, the ratio between the

initial animal weight w_0 and $w = \left(\dfrac{a_1}{a_2} \cdot \dfrac{b_1 - 1}{b_2 - 1} \right)^{\frac{1}{b_2 - b_1}}$ has great significance; accordingly, the specific growth rate may or may not increase during the postembryonic period. These conclusions are strictly theoretical and require practical verification.

Not long ago the work of Kruger (1968) analyzing growth in the medusa R h i z o s t o m a   o c t o p u s   came to my attention. He found that the specific growth rate in this species increases with age and on this basis, he established a peculiar type of growth. Kruger's

growth equation $w = mn^t$ fits satisfactorily the growth of this particular medusa, but cannot have general application; it remains a special mathematical expression referring to a particular case. At the same time, the data on the growth of R . o c t o p u s   indicate that the type of growth which I have suggested, in which the specific growth rate accelerates with age, indeed exists in nature.

Unfortunately, detailed information on the growth of other medusas is not available to me. Mironov (1967) prepared a growth curve of A u r e l i a   a u r i t a   based on incomplete data. Significantly, the concave part of this curve also indicates an increase in the specific growth rate with age.

At present, equation (37) cannot be used directly for describing the **growth curve** of any animal, since no work has been found in which the quantity $A - \dfrac{dw}{dt}$ is represented as a function of weight $\left(A - \dfrac{dw}{dt} \right.$ represents the expenditure for respiration together with vital losses of matter). With respect to some crustaceans there are data allowing construction of an individual **production curve** from equation (37). Let us consider, for example, the elements of the energy balance of A r t e m i a   s a l i n a   (Sushchenya and Khmeleva, 1967; Khmeleva, 1968). All the quantities necessary for the calculation are expressed here in grams of dry weight and refer to a temperature of 25°C. The following conversion factors were used for this purpose: dry/fresh weight ratio of 0.2 in A r t e m i a   s a l i n a , 0.13 in the food organisms (the alga D u n a l i e l l a); the ration determined at 20° (at a high food concentration) was predicted for 25° according to the Krogh curve (Vinberg, 1956); assimilability was taken as 0.8 (Khmeleva, 1968, assumes an assimilability of 0.5).

Hence, equation (37) for Artemia assumes the following form:

$$\frac{dP}{dt} = 0.031w^{0.705} - 0.012w^{0.688}. \tag{42}$$

When integrating this equation it is necessary to take into account what was mentioned at the beginning of the present chapter, namely, that in calculating the production curve the quantity ΔP depends on w but not on the attained value of P. Having calculated ΔP for a certain time period Δt, we cannot assume that the weight has also changed by ΔP. Since part of the matter is eliminated during life, the change in weight as a function of age has a different pattern which must be known in advance in this case.

According to Khmeleva (1968) the weight gain of Artemia is adequately described by the expression

$$w_t = (0.090 - 0.076e^{-0.063t})^3 \tag{43}$$

(weight in grams dry matter, temperature 25°C). This equation is applicable up to the age of 30 days when growth ceases at an actual definitive weight of 0.00045 g.

In relation to this, when integrating equation (42), w must be expressed as a function of time (age) according to equation (43). Thus, the production curve of Artemia up to one month of age is obtained. Since the weight of these animals does not change later, $\frac{dP}{dt}$ remains constant (as seen from equation (42)). Therefore, a linear increase in production for Artemia older than one month was assumed in which $\frac{dP}{dt}$ corresponds to a weight of 0.00045 g.

The production and growth curves of Artemia salina are shown in Figure 6. These curves closely resemble those obtained for Daphnia on the basis of factual data (see Figure 3). According to Khmeleva (1968), a female Artemia produces progeny having a total dry weight of 0.0033 g during 130 days of life, and all the larval molts weigh 0.00058 g. The total of these values and the definitive weight of the female (0.00045 g) equals 0.00433 g, which is the individual production during 130 days, estimated using the "output" of the system. The production curve for 130 days yields a theoretical production of 0.008 g, which is double the estimated individual production. From available data it is impossible to determine which of these values is closer to reality. On the one hand, calculation from the production curve is not very accurate because the assimilability coefficient was only approximate; more-over, the equations relating to the ration and metabolic losses as a function of the body weight are not quantitatively reliable (the ration equation, for example, refers to another temperature). According to Khmeleva, nauplii do not consume food during the first 3—4 days of life. Even minor changes in the constants of equation (42) can alter the production curve. On the other hand, it must be assumed that the sum of the definitive weight and those of progeny and larval molts does not take into account that part of the individual production including liquids secreted during life.

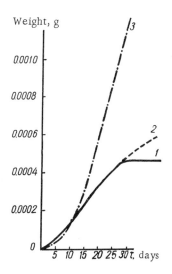

FIGURE 6. Growth and individual production of Artemia salina:

1) weight increase according to experimental data; 2) weight increase according to equation (43) (1, 2 after Khmeleva, 1968); 3) individual production according to equation (42) (all values are expressed in grams dry weight).

On the whole, available factual data are not sufficient for an effective exploitation of equation (37). In thorough studies of the energy balance, however, this equation is useful for comparing such quantities as food consumption, assimilation, metabolic losses, growth and production on the individual level.

Chapter III

METHODS FOR CALCULATING PRODUCTION
OF ANIMAL POPULATIONS

1. "CALCULATION" AND "DIRECT" METHODS
FOR STUDYING PRODUCTION

Examination of works on bioproductivity (including works on
primary and bacterial production) shows that there are two fundamental approaches for evaluating production indexes. First,
production can be determined on the basis of various combinations
of physiological and ecological data (in particular, individual growth,
and reproduction rates, age structure of the population, population
and biomass dynamics, elimination rate), which individually do not provide any information on production. Consequently, starting from these
initial data production can be calculated by various calculation
schemes. This method of studying production will be referred to
as the calculation method.

Second, ways are being sought for direct measurement of the
rate of synthesis of organic matter or of quantities proportional to
it. In this case, the evaluation of the production rate also involves
calculation, but here it is relatively simple. Accordingly, this
approach can be regarded as the **direct method** for studying production. Intermediate cases are possible, since the boundary
between the two methods is not absolute. I shall attempt briefly to
point out the characteristic features of each method and of the
results achieved by their means.

Scientists interested in a given quantity attempt to measure it
instrumentally. Calculated estimates of this quantity are usually
considered second best. I shall not enter into a general discussion
of this topic, but for production research it can be said that direct
determination of the production is preferable to the tedious collection of ecological and physiological data and the subsequent complex
calculation of production. The advantages of the direct method for
production studies are evident in the case of work on primary
production. The development of relatively simple and rapid direct
procedures for measuring the production of photosynthetic plants

(especially, the radiocarbon method) has led to the rapid accumulation of data on primary production. These methods allowed for estimates of the production of the phytoplankton as a whole, i. e., the production of an entire trophic level comprising a great number of populations. Production estimates by this method characteristically do not require information on the species composition, proportions of numbers and biomasses of different species or the reproduction rate of the algae.

However, production research is not completed with the production estimate. It is also necessary to **explain** the recorded values and **predict** them for different conditions. Here, too, it is evident that the value of all phytoplankton production should be supplemented with assessment of the roles of the individual species constituting the community. As a result, workers have engaged in field studies of primary production using the radiocarbon method. They have separated the phytoplankton into size groups, prior to exposure in the flasks, in order to evaluate the roles of the different fractions. Simultaneously, laboratory research on algal physiology has intensified, especially with reference to reproduction and photosynthetic rates of different populations under different conditions such as temperature, illumination, supply of biogenic elements, etc. Finally, workers at the Institute of Biology of Southern Seas of the Academy of Sciences of the Ukrainian SSR (Ten, 1964; Kondrat'eva, 1965, 1968; and others) have developed a calculation method for primary production in the sea, which is used alongside the direct method, although it is more tedious.

Thus, the calculation method for production studies has its own advantages; it clearly reveals the relationship between the basic ecological and physiological parameters of the population or community. As a result, this facilitates explaining and forecasting the productivity of a system under certain conditions. It is also important that production research based on the calculation method requires knowledge of many parameters, which exert a great stimulatory and guiding influence on ecological and physiological studies.

At present, phytoplankton, bacterial and zooplankton productivities are studied in manifold ways. In the area of primary production, as already mentioned, direct methods predominate for studying production of all trophic levels of producers as a whole. However, there is a tendency to study the productivity of separate species of the populations (by calculation and direct methods). These trends will probably intensify in the future.

In most studies bacterial productivity is usually assessed as a whole or for the entire bacterioplankton, mainly by calculation procedures. Direct methods for determining bacterial production are being developed at present, using once again the total production of bacterioplankton (radiocarbon method).

So far zooplankton productivity is assessed by calculation only, and strictly for species populations (not taking into account approximation of the production of the total zooplankton). Knowledge of the productivity of common species leads to an estimate of the production of the community. The situation is more fortunate with respect to the productivity of invertebrates, since any problem related to their productivity can be analyzed more thoroughly. This is due to the wide application of calculation methods for assessing productivity in different species based on a thorough knowledge of the ecological and physiological parameters of the population. Of course, a search for direct methods for determination of animal production is required (Chmyr, 1967; Shushkina and Sorokin, 1969), but calculation methods have their own value and should not be completely replaced by direct ones.

Having outlined the two basic approaches to production determination, it must be pointed out that the rest of this chapter will be devoted strictly to calculation methods and that the body of factual material in this work was obtained by indirect methods.

2. BASIC SCHEMES FOR CALCULATING PRODUCTION INDEXES

According to the calculation method for studying productivity, production indexes are estimated on the basis of a complex of physiological and ecological data which individually could not completely characterize productivity. Strictly speaking, collection of these initial data does not belong to the realm of production research. For example, data on animal growth are widely used in production calculations, although growth research is usually unrelated to productivity problems. To quote Vinberg (1936, p. 598), "the study of the basic physiological features of individual organisms, essential for an understanding of the principles of productivity, does not in itself constitute research in productivity."

Thus, productivity research consists strictly either in the derivation of production indexes from a complex of preliminary data or in the direct measurement of these indexes. During productivity research, however, a scarcity or lack of the necessary preliminary data on the population examined may necessitate the collection of the necessary information. For this reason, productivity research includes broadly the collection of production features together with initial data. This approach is represented

in the handbook "Methods for the Determination of Production ..."
(Vinberg, 1968 a) where methods for weighing animals and determining caloric content, growth rate, etc., occupy about 40% of the
book. This was justified and highly useful, since information contained in this single book aids in determining the production of a
population with unknown characteristics. In the following discussion of production calculation methods, it has been assumed that
the necessary preliminary data are available and were obtained
by reliable procedures.

In working out actual procedures for production calculations
the following topics are of particular importance: a) the availability
of preliminary data on the examined population; b) the production
indexes to be evaluated; c) the necessary degree of accuracy in
the calculation.

The many possible solutions provide a number of useful variants
for calculating production indexes. Most of the calculation variants
proposed so far have appeared in the handbook (Vinberg, 1968 a).
Here an attempt will be made to classify these variants into basic
types of calculation schemes.

The number of these variants depends upon the desired degree
of generalization and the choice of classification principles. The
classification proposed below is clearly not the only possible one.
The following discussion of the expediency of calculation schemes
for production rests on differences in the character of preliminary
data used in the calculation, as well as on principles explained in
Chapter I.

First of all, production calculations can be classified as based on
"input" or "output" of the system (see Figure 1). The main difference between these categories with respect to preliminary data is
that only "output" calculations utilize values of eliminated biomass.
On this basis, calculations based on the "output" of the system
are treated as a separate scheme (first scheme).

For calculating production according to the input of the system,
a variety of preliminary data are used. In particular, there is the
calculation of production as the difference between assimilation and
losses (second scheme), whereas in other cases data on the increase
of the biomass of the population are used in calculation.

Among the latter cases, calculations vary with respect to the path
method of conversion from population size to biomass. For this the
relationship between individual weight and age can be used. This
procedure yields more detailed biomass estimates, but inevitably
involves data on the age structure of the population and the relationship between the individual growth rate and weight (third scheme).
In other cases the biomass is estimated only according to population
size and average individual weight. For this purpose, average

indexes are used, presumably characterizing the whole population, such as the average individual weight and average reproduction rate per individual. Biased estimates are possible here since the average indexes calculated for a given state of the population may differ from actual averages because of changes in the age structure occurring in the time period during which production is determined. The possible discrepancy in the calculation is insignificant for shorter life cycles and for smaller differences between minimal and maximal individual weights. This way of calculating production will be referred to as the one based on population dynamics (fourth scheme).

First scheme for calculating production. This scheme involves production calculations based on the output of the system, where production is determined as the change in existing biomass plus eliminated biomass. This is the approach of Boysen-Jensen (1919), among others. The calculation is made according to equation (1):

$$P_t = B_t - B_0 + B_e.$$

Several calculation patterns exist within this scheme. For example, production equals eliminated biomass if the existing biomass remains unchanged during the test period.

Calculations according to this scheme require the following preliminary data: 1) initial biomass; 2) final biomass; 3) biomass eliminated during test period. In the case where eliminated biomass is negligible, production equals the change in existing biomass.

Second scheme for calculating production. Clarke et al. (1946) proposed estimating production as the difference between the rate of food assimilation by the population and the rate of metabolic losses. This procedure is only rarely used because it is rather difficult to determine the rate of total assimilation of food in the population; but it has importance in a general sense since it provides evidence for the theoretical analysis of the production process in systems of different complexity from the individual to the biocenosis (see Chapters I and II).

Third scheme for calculating production. This scheme includes all procedures based on data relating to individual growth and the age structure of the population. Currently this is the most widely used scheme. Konstantinov (1960) proposed calculating production according to the rate of the weight increment in all individuals of the population. Later variants of this scheme also calculate production by reproduction (Pechen' and Shushkina, 1964; Greze and Baldina, 1964). Production is determined by the equation $P = P_1 + P_2$. The numerous variants employed reflect different methods for determining the individual increment.

Preliminary data are 1) weight increments of individuals of different ages; 2) age structure of the population; 3) weight of progeny (unless included in the weight increment of the females); 4) average biomass of the population.

Fourth scheme for calculating production. This scheme is based on data on population dynamics. To avoid analysis of individual growth and age structure, all individuals are regarded as having a certain average weight. Then production can be determined according to the reproduction rate. The "turnover rate" of the population size (reciprocal of the doubling time of the population) is frequently used.

Preliminary data are 1) reproduction rate; 2) average size of the population; 3) average individual weight.

The main procedures for calculating the production of populations have been classified to guide the reader through the many calculation variants, most of which are described in detail in the handbook "Methods for the Determination of the Production ..." (Vinberg, 1968 a). The following sections of this chapter contain results of original research.

3. THE "PHYSIOLOGICAL" METHOD FOR CALCULATING PRODUCTION

Under this title, the handbook (Vinberg, 1968 a) presents an approximate calculation variant belonging to the third scheme and based on the following preliminary data: 1) population biomass; 2) age structure; 3) relationship between metabolic rate and weight; 4) caloric content of the animals; 5) the values of K_2 ($K_2 = \dfrac{dw}{dt} \cdot \dfrac{1}{A}$, i. e., the ratio between increment and assimilated food).

Equations have been worked out giving the increment for animals of different weights on the basis of these data and allowing calculation of production. These equations were found to apply to animals with almost parabolic growth. This method is based on the relationship between animal growth and metabolism, established by Vinberg (1966) and has been used for specific calculations in a number of cases (Vinberg, 1968 a; Shushkina, 1968). The values obtained by this method roughly correspond to those assumed a priori for the respective species, but some of the basic assumptions of the method require further analysis and clarification.

Vinberg (1966) proceeds from the well established relationship between respiratory losses and body weight:

$$T = a_2 w^{b_2}. \tag{44}$$

Assuming that in some cases the ratio v between the increment $\frac{dw}{dt}$ and the losses remains constant during the period of individual growth,

$$\frac{dw}{dt} \cdot \frac{1}{T} = v = \text{const,} \qquad (45)$$

we obtain $\frac{dw}{dt} = vT$, and according to equation (44)

$$\frac{dw}{dt} = v a_2 w^{b_2}. \qquad (46)$$

Thus Vinberg (1966) obtained the parabolic growth equation in differential form.

Let us denote $n = 1 - b_2$ and assume that $w_0 = 0$. In that case the individual weight from equation (46) can be expressed as a function of the age τ:

$$w = (v a_2 n \tau)^{\frac{1}{n}}. \qquad (47)$$

Since the condition $v = \text{const}$ yields an equation of parabolic growth, the converse conclusion can be reached that "... when growth is parabolic, v and K_2 are constants" (Vinberg, 1968 a). The physiological method of calculation for animals with parabolic-type growth is based precisely on the conclusion that v and K_2 are constants in such animals.

This problem has been examined from a mathematical viewpoint (Zaika and Makarova, 1969). The very fact that $v = \text{const}$ leads to parabolic growth does not necessarily justify the opposite statement. Indeed, there are functions $v = f(\tau)$ in which v decreases while growth remains parabolic. For example, v can be a power function of τ:

$$v = u \tau^{m-1}, \qquad (48)$$

where u and m are constants.

From equations (47) and (48) we obtain that $w = (u a_2 n \tau^{m-1} \tau)^{\frac{1}{n}}$, or, setting $z = u a_2 n$:

$$w = z \tau^{\frac{m}{n}}. \qquad (49)$$

According to equation (48), if $\frac{m}{n} > 1$ we have a parabolic growth, and v decreases with increasing τ. In this case the specific rate of weight increase is

$$q_w = \frac{m}{n} \cdot \frac{1}{\tau},$$

i.e., q_w also decreases with age.

In parabolic growth, K_2 may theoretically remain constant during growth, but this is not necessarily so. This problem must be solved experimentally. According to Petipa (1966 a, b), growth in the copepods C a l a n u s   h e l g o l a n d i c u s and A c a r t i a   c l a u s i is near-parabolic. However, K_2 in the former species increases during the nauplius stage and then steadily declines, whereas in the latter species this quantity seems to be nearly constant. This lends practical support to the theoretical conclusion that K_2 may change during parabolic growth, but wider, more reliable data are still necessary.

This limits the application of the physiological method of production calculation. Indeed, suppose we wish to estimate the production of an animal presumably growing parabolically. Having found the value of K_2 for a given weight category, the physiological method cannot be used since it is uncertain that K_2 is a constant. If K_2 is investigated for all age groups, then the physiological method for calculating production is not necessary because methods for simultaneous determination of K_2 and the increment do not exist. In other words, if K_2 is known for all age groups, the values of the increments are also known, and production can be calculated by conventional procedures.

4. METHODS FOR CALCULATING PRODUCTION
OF MICROORGANISMS

The term "microorganisms" is used here broadly to cover bacteria, unicellular algae and protozoans. This complex of organisms in the plankton was called "protoplankton" by Wood (1965).

Such groups as radiolarians, foraminifers, small heterotrophic flagellates and infusorians have not been studied in terms of productivity. Among the few exceptions are the works of Shcherbakov (1963), Shushkina (1966) and some others who assumed tentative values for the specific production of infusorians in evaluating production of lake zooplankton. On the other hand, there is a vast literature dealing with the productivity of photosynthetic algae, i.e., primary production. This topic, however, is outside the scope of this work.

The concept "production," worked out for multicellular animals, applies as well to microorganisms. In the latter case, however, production determinations cannot be based on indexes such as individual weight growth or size (age) structure of the population because of the minute size and short individual life cycle of micro-organisms. It is much simpler to assume that all the cells have a certain average weight, and to calculate the mass of all production in terms of reproduction rather than as the individual increment. This does not mean, however, that the production of microorganisms can be determined only according to the fourth scheme (see Chapter III); equation (1) can serve satisfactorily for this purpose.

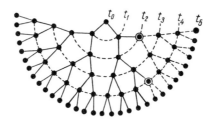

FIGURE 7. Scheme of the formation of a bac-terial colony. The eliminated individuals are encircled: $t_0 - t_5$ are consecutive time points at one-hour intervals.

To facilitate the further discussion of this topic we shall con-sider an ideal situation for the formation of a microcolony by reproduction (Figure 7). We shall assume that each cell has unit biomass and division occurs synchronously each hour, i. e., a generation time of $g = 1$. Eliminated cells in the scheme are encircled. The production at $t = 5$ hours can be calculated as follows. It is evident from the scheme that $B_0 = 1$, $B_t = 22$, $B_e = 2$. According to equation (1) we obtain $P_t = 22 - 1 + 2 = 23$. Determina-tion of production of the colony at hourly intervals leads to the conclusion that growth is irregular. This is because the total mass of the colony increases stepwise, while the individual reproduction rate remains constant. Indeed, production can be expressed as $P = CB$. The specific production C in this model is constant because it depends on the reproduction rate. However, production per unit time increases with the biomass. If the production during any hour interval is divided by the initial biomass for the same interval, then it is evident that C equals unity in all cases (this is a simplification since, as will be shown below, the actual value of C is 0.693 because the production must be divided by the average biomass). However, the hourly specific production cannot be determined by dividing the production for several hours by the initial biomass and then dividing the obtained result by the time, i. e.,

$$C = \frac{P_t}{B_0 t} .$$

Here the initial biomass is obviously unsuitable for calculation of the average specific production. Indeed, a different result for each value of t is obtained by this equation.

The method of Ivanov (1955) for calculating bacterial production has been adopted by many authors (Novozhilova, 1955, 1957; Salmanov, 1959; Krasheninnikova, 1960; Gambaryan, 1960; Drabkova, 1965). This method is described in practical manuals (Kuznetsov and Romanenko, 1963; Rodina, 1965). Ivanov describes the procedure thus. Let us start with a bacterial population whose initial biomass B_0 and generation time are known. The proportion of the total bacterial mass dividing within one hour equals $\frac{B_0}{g}$. Consequently, the bacterial increment in t hours without elimination will be $\frac{B_0 t}{g}$ and the bacterial mass will be $B_0 + \frac{B_0 t}{g}$ by the end of this time. With elimination by consumption equal to B_e bacteria per hour, the biomass of the population after t hours will be $B_t = B_0 + \frac{B_0 t}{g} - B_e t$. Hence

$$B_e = \frac{B_0 - B_t}{t} + \frac{B_0}{g} .$$

According to Ivanov, this is the mathematical expression of hourly production. However, it is obvious that Ivanov identifies production with consumption. To convert the value for consumption to that of production it is necessary to add the difference between the final and initial biomass. This was first noted by V. V. Menshutkin (Kozhova, 1964). Moreover, it was shown (Kozhova, 1964; Romanova and Zonov, 1964) that the calculation must be based on geometric rather than linear bacterial growth. Consequently, the consumption equation becomes

$$B_e = \frac{b \, (B_0 e^{bt} - B_t)}{e^{bt} - 1} , \qquad (50)$$

where $b = \frac{\ln 2}{g}$ is the reproduction ratio (see Chapter I).

Unaware of Kozhova's work (1964), I also analyzed in detail the sources of error in Ivanov's method (Ivanov, 1955) and indicated that it is necessary to proceed from equation (1), in which the value of hourly consumption can be derived from equation (50) if experimental conditions raise no doubt about the respective assumptions on the character of elimination (Zaika, 1967b). The variants for calculating bacterial production, developed by Ierusalimskii (1949, 1954, 1963) for data obtained from overgrown slides exposed in water, were analyzed. Ierusalimskii bases himself on the value of the average specific production rate, which is expressed by the following equation for the time period t:

$$\bar{C} = \frac{1}{t} \ln \frac{B_t}{B_0}, \quad \text{or} \quad \bar{C} = \frac{2,3}{t} \log \frac{B_t}{B_0}. \tag{51}$$

This is indeed the quantity C mentioned above, since

$$P = \frac{dB}{dt}, \quad C = \frac{P}{B} = \frac{dB}{dt} \cdot \frac{1}{B} = \frac{d(\ln B)}{dB} \cdot \frac{dB}{dt} = \frac{d(\ln B)}{dt},$$

whence equation (51) for the average specific production rate is derived. As shown earlier, however, this equation yields a correct value of C only when elimination during the test period equals zero. It should be noted that the more elimination differs from zero the smaller the value of the specific production rate derived from equation (51). To obtain the specific production rate in the presence of elimination, it is necessary to add the specific elimination $\frac{B_e}{B}$ to the right side of equation (51):

$$C = \frac{1}{t} \ln \frac{B_t}{B_0} + \frac{B_e}{B}. \tag{52}$$

If one accepts, as did Ivanov (1955), a given hypothesis on the nature of elimination, (i.e., elimination results from consumption which proceeds at a constant rate in experimental conditions), the production during the experiment can be calculated as follows. If $B_e = 0$ during the time t:

$$P_t = \overline{BC}t = \bar{B} \ln \frac{B_t}{B_0},$$

where $\bar{B}$ is the average biomass for the time t. If $B_e \neq 0$ during the time t:

$$P_t = \bar{B} \ln \frac{B_t}{B_0} + B_e t, \tag{53}$$

where B_e is given by equation (50).

There are two corrected variants for calculating production: equation (1) in the form $P_t = B_t - B_0 + B_e t$ and equation (53). In both cases B_e can be determined from equation (50). Verification with actual data shows that both methods yield practically identical results (Zaika, 1967 b).

According to Kozhova (1964), under experimental conditions preventing consumption, the difference between final and initial amounts of bacteria equals production. The production in a test flask under such conditions exceeds that obtained in the same period from a similar volume of water in nature. Without elimination, the bacterial biomass will increase during the experiment, whereas in natural waters it remains stable or increases more slowly than in the flask. Thus, the average bacterial biomass in a flask must exceed that obtained in the water body, i. e., the production must be greater in the flask. The specific production determined in such laboratory experiments can be used later for calculating production in a water body with respect to the actual biomass.

The example of bacteria has been used for analyzing some possible variants in calculating production of microorganisms. Although seemingly different, the methods of Ivanov and Ierusalimskii produce identical results after being put in a more accurate form. These considerations are not limited to bacteria, but apply also to other microorganisms. For example, Ten (1964) worked out a method for calculating production of unicellular algae based on cell reproduction and elimination in flasks. This procedure closely resembles the Ivanov method (in modified form).

The variants enumerated for calculating production involve an analysis of changes in the existing biomass and their elimination, i. e., they belong to the first calculation scheme (see Chapter II). A study of the relationship between the biomass increment and elimination might be interesting. However, the fourth calculation scheme is preferable if the only value sought is production. In the latter case the preliminary information required is minimal.

It was shown in Chapter I that, regardless of whether elimination occurs or not, the specific production C of a population equals reproduction coefficient b. The significance of this coefficient will be explained in greater detail using the model of exponential increase of the number of microorganisms.

Analyzing a species population of microorganisms, it is assumed for simplicity that all the cells are dividing at an equal (average) rate. This means that, without elimination, the rate of biomass increase is proportional to the existing biomass, and the proportionality coefficient is equal to the reproduction coefficient b. Thus, without elimination

$$\frac{dB}{dt} = bB,$$

$$B_t = B_0 e^{bt}.$$

From equation (51) it follows that without elimination $C = b$:

$$C = \frac{1}{t} \ln \frac{B_t}{B_0} = \frac{1}{t} \ln \frac{B_0 e^{bt}}{B_0} = \frac{1}{t} \ln e^{bt} = b.$$

There is a definite relation between generation time (g) and the reproduction coefficient in organisms reproducing by binary fission:

$$b = \frac{\ln 2}{g}.$$

This yields a simple method for calculating specific production:

$$C = \frac{\ln 2}{g} = \frac{0.693}{g} \,. \tag{54}$$

Obviously, for determination of C it is only necessary to know the generation time. As for production, its determination requires knowledge of the population biomass.

In developing this method (Zaika and Makarova, 1970) we devoted great attention to the effect of the value of elimination on the specific production calculated according to the equation

$$C = e^b - 1. \tag{55}$$

This effect can be defined by expressing the specific production in the following form:

$$C = \frac{P}{B_0} \,,$$

where $P = B_t - B_0 = B_0 (e^{bt} - 1)$ and $t = 1$. Equation (55) as shown by Ten (1964) refers to cases without elimination. If it is assumed that elimination does not equal zero, it becomes clear that this quantity influences specific production; in other words, equation (55) yields an excessively high value of C with elimination. In our work, cited above, we attempted to prove that ignoring elimination produces only a negligible error. Unfortunately, we did not realize that all the calculations were unnecessary if production is divided by the average biomass (the correct approach) rather than by the initial biomass for the given hour.

In other words, we obtained by different means two equations for specific production of microorganisms, namely, equations (54) and (55), and erroneously supposed them to be equivalent. The correct equation is (54), since elimination need not be taken into account when specific production is calculated by this method.

5. METHODS FOR CALCULATING PRODUCTION OF MULTICELLULAR ANIMALS BASED ON DATA ON POPULATION DYNAMICS

Possibilities of applying the fourth scheme to calculating the production of a population have been discussed generally in Chapters I and II, and in Chapter III with reference to micro-organisms. This scheme has been applied to multicellular animals in different variants whose affinities are not always evident. As already noted, determination of production on the basis of population dynamics can yield somewhat incorrect results because changes in the age composition of the population are not accounted for. Consequently, the relevant methods are only briefly described in the handbook (Vinberg, 1968 a), and not mentioned at all

by Shpet (1962, 1964, 1965) and Pidgaiko (1965) in their work on
"potential production."

Variants of the fourth scheme for calculating production of
multicellular organisms have been used outside the Soviet Union
mainly in the United States. In the relatively early works of Juday
(1940, 1943), Lindeman (1942) and others, zooplankton production was
determined by multiplying average annual biomass by the number of
turnovers per year, which is expressed as $\frac{365}{T}$, where T represents
turnover time in days and is considered to equal the average life
span of the particular animals. The reciprocal to turnover time
$\left(\frac{1}{T}\right)$ is the "turnover rate" which provides an approximation of the
production rate:

$$C \approx \frac{1}{T} . \tag{56}$$

Later determinations of the turnover rate were based on a more
thorough analysis of population dynamics, rather than on average
life span alone. Works using this method (Edmondson, 1960; Stross,
Nees and Hasler, 1961; Hall, 1964; Wright, 1965) are discussed in
the handbook (Vinberg, 1968 a) in which the following comment is
made: "This procedure for calculating the production applies only
to a stationary population with equal mortality in the age groups,
i. e., populations with a constant age composition. Otherwise the
turnover time of population size differs from that of the biomass; thus
this apparently simple method becomes inapplicable."

In the light of this judgment and since the age composition of
any population cannot conceivably remain constant during any long
time period, the application of the fourth scheme to calculating production of multicellular animals is practically ruled out. However,
I believe that this judgment is unnecessarily severe. Changes in
age structure undoubtedly affect the average individual weight and
average reproduction rate which in its turn influences the turnover
rate. However, for determining production on the basis of population dynamics it is necessary to use indexes calculated for a sufficiently typical condition, i. e., for some average state of the given
population. If this condition is maintained, the turnover rate can
rather satisfactorily reflect the average specific production of the
population. In practice, the average value of the turnover rate must
be calculated for a period several times longer than the turnover
time. As already noted, calculations based on population dynamics are more accurate if the difference between initial and final
weights of the animals is smaller. It is noteworthy that
procedures for calculating production of rotifers described in

the handbook (Vinberg, 1968 a) are variants of the fourth scheme and this does not detract from their value.

The scheme for calculating production based on population dynamics is of theoretical interest because it reveals the relationship between the problems of productivity and population dynamics. Certain publications outside the Soviet Union have described the determination of the coefficient r for "intrinsic rate of increase." Its application will be included in the following discussion. In the first chapter it was shown that

$$r = b - m \tag{11}$$

and

$$C = b, \tag{13}$$

where b is the reproduction coefficient and m the elimination coefficient. If elimination is relatively small, it can be assumed that $b \gg m$, $C \approx r$. Thus, in certain circumstances r provides a minimal estimate of the average specific production rate (this topic is discussed also in Chapter VI).

METHODS FOR CALCULATING THE SPECIFIC PRODUCTION OF ROTIFERS

Methods for determining production of rotifers (Galkovskaya, 1963, 1968) can illustrate the application of the fourth scheme for determining production of multicellular animals. Rotifers are smal animals with a brief life span; the different age groups hardly differ morphologically. These properties make them highly suitable as test items for the fourth calculation scheme, which is based on population dynamics data.

Galkovskaya (1968) describes two variants for determining the production of rotifers. The first is based on the equation proposed by Edmondson (1960) for calculating the average number y_1 of eggs deposited per female in the population during 24 hours:

$$y_1 = \frac{F}{ND_1}, \tag{57}$$

where F is the number of eggs in the sample, N the number of females and D_1 the duration of embryogenesis. Knowing the value of y_1, the production of the population can be calculated according to the equation

$$P_t = y_1 \bar{N} t, \qquad (58)$$

where P_t is the production of all individuals during time t and $\bar{N}$ is the average number of females during the period t.

The second procedure differs from the first in that y_1 of equation (58) is replaced by the expression

$$y_2 = \frac{1}{T'}, \qquad (59)$$

where T' is the doubling time of numbers and y_2 is its reciprocal. The doubling time is defined as the period from hatching of a rotifer to hatching of its progeny. The number $\bar{N}$ obviously reflects the average number of individuals in the population, regardless of age.

With respect to the first procedure it can be said that if $\Delta t = 1$

$$y_1 = \frac{\Delta N}{\Delta t} \cdot \frac{1}{N_0},$$

and if $\Delta t \to 0$

$$\frac{dN}{dt} \cdot \frac{1}{N} = \ln (y_1 + 1).$$

Therefore,

$$C = \ln (y_1 + 1)$$

(provided that all data are expressed in terms of population size).

The second procedure resembles the method described in Chapter III for determining production of microorganisms. It is evident from equation (58) that the index y_1 (or y_2) must indicate specific production. No other meaning is possible. It is seen from equation (59) that Galkovskaya considers specific production as the reciprocal of doubling time. The generation time g of microorganisms can also be regarded as the doubling time. However, we know that the specific production of the population is related to this index, thus

$$C = \frac{0.693}{g}. \qquad (54)$$

Consequently, $C = \frac{0.693}{T'}$ rather than $C = \frac{1}{T'}$ must be used in

equation (58) for rotifers, as well as in other cases. It should be noted that Galkovskaya, like Ivanov (1955), based her calculations on a linear model of population growth, although the exponential model yielding equation (54) for specific productions is more correct.

"POTENTIAL PRODUCTION" OF ANIMALS

Finally, it would be pertinent with respect to production determinations based on population dynamics to consider research on "potential production" of animals (Shpet, 1962, 1964, 1965, 1968; Pidgailo, 1965, 1968). These studies depart somewhat from the mainstream of production research because their exact relation to conventional determinations of production is uncertain (this applies to theoretical aspects, since empirical comparisons of production and potential production values have been made). Shpet and Pidgaiko determined animal productivity using specific reproduction progressions showing a theoretical increase in numbers from a single individual or parental pair. This approach applies to population increase in ideal conditions, i.e., in the absence of predators, food shortage, etc. Losses are attributed entirely to natural mortality. However, individual reproduction rates and life span are determined from actual data obtained experimentally or in field observations. Although the increase in numbers is derived abstractly, the reproduction rate reflects rather accurately the actual state under favorable conditions.

According to this method, productivity is defined as the biomass reached after a time interval from the moment of existence of a single individual or pair. This approach is not quite satisfactory since absolute magnitudes obtained from different animals cannot be compared, especially since the test period is defined arbitrarily and differs among animal groups. At any rate, a comparison of the fourth scheme leads to the conclusion that a correctly formulated reproduction progression readily yields the reproduction coefficient, and consequently the specific production of the population. Thus, data on potential production can easily be made relevant, and approximations obtained from it are fully comparable with the other data on animal production. This can be done using the equation

$$C = \frac{\ln B_2 - \ln B_1}{t_2 - t_1},$$

which applies to the advanced part of the reproduction progression, where the "age structure" is sufficiently complex. It is unfortunate that Shpet (1968) presented complete reproduction progressions for only a few species and indicated only the final biomass in most cases.

Thus, Shpet (1968) shows in Table 7 the weight of a single individual and the biomass reached by different planktonic animals after 36 days. Therefore, the procedure proposed here is as follows. The individual weight will be regarded as the initial biomass B_0 used in the above-mentioned equation

$$C = \frac{\ln B_t - \ln B_0}{36},$$

where B_t is the biomass after 36 days.

Using this method, we obtained the following values for C for some of the species listed in Table 7.

Species	c (per 24 hours)
Daphnia magna	0.5
Moina sp.	0.51
Daphnia longispina	0.3
Ceriodaphnia reticulata	0.27
Acathocyclops viridis	0.14
Mesocyclops leuckarti	0.17
Brachionus rubens	0.33

A comparison of these values with those calculated for the same species by other methods shows (see Chapter V) that these production estimates are quite satisfactory. (It is noteworthy that Shpet has used different initial data; moreover, B_0 had to be taken as the weight of a single individual rather than the biomass at a time when the population has a sufficiently complex age composition.) However, even this peculiar variant of the fourth scheme yields estimates quite acceptable for approximations of production.

Chapter IV

PRODUCTIVITY OF INFUSORIANS

The research on the productivity of heterotrophic micro-organisms has only begun. In Chapter III the practical problems associated with this topic were discussed and a simple procedure described for evaluating the production of microorganisms on the basis of their biomass and reproduction rate. Matters are complicated by the fact that scant preliminary data, necessary for production calculations, are available and are not always reliable. The present chapter summarizes available information on abundance and reproduction rates of various species of infusorians. A rough estimate of the productive capacities of this important group of the heterotrophic microplankton in conditions of certain biotopes will be based on these data.

1. QUANTITY OF INFUSORIANS IN PLANKTON

Quantitative studies of infusorians are not numerous, although Lohmann (1908) demonstrated long ago that infusorians constitute 10 — 20% of all metazoans in the Kiel fjord. The scarcity of information on the infusorians can apparently be attributed to the neglect of the role of microorganisms in aquatic communities and to the lack of simple and reliable means for collecting quantitative samples of microplankton. Standard methods for sampling plankton and benthos do not provide a complete idea of the abundance of infusorians.

The marked recent trend toward studying major ecosystems as functional entities, with particular attention to the role of all the principal elements, explains the publication of censuses of infusorians in water bodies. Works on freshwater infusorians have been published by Shcherbakov (1963, 1969), Mordukhai-Boltovskaya (1965) and Eggert (1967, 1968). Marine planktonic infusorians are discussed in the articles of Margalef (1963 b, 1968), Beers and Stewart (1967) and Grøntved (1962). The early work of Lackey

(1936) and the extensive studies of Fenchel (1967, 1968) and others deal with marine benthic infusorians. Some data are available on the abundance of infusorians in sewage and psammon. Due to the lack of reliable field procedures, the different authors have used a large variety of collection, fixing and treatment methods.

TABLE 1. Density of infusorians in some marine water bodies

Water body	Density of infusorians, organisms/liter	Source
Pacific Ocean near California	10–80	Beers and Stewart, 1967
Ionian and Ligurian seas (open sea)	2–80	Zaika, 1969
Neritic zone of the Black Sea	more than 5–20	Zaika, 1970 e
Coastal waters of the Mediterranean Sea	50–1,000	Margalef, 1963 b, 1968
Sevastopol Bay, Black Sea	2,000–4,000	Zaika and Averina, 1968

It must be noted that, like many microorganisms, infusorians show considerable changes in abundance corresponding to brief and comparatively small environmental changes. Hence it has often been noted that major fluctuations in numbers of infusorians occur in all biotopes. A relation has been discovered between average density of infusorians in marine plankton and distance from shore, as illustrated by averaged data presented in Table 1. The values given are far from maximal. Thus, some samples obtained by Margalef (1963 b, 1968) contained as much as 13,000 infusorians per liter. Such densities are due to sharply localized peaks of reproduction, especially in areas rich in "fresh" detritus (Zaika, 1969 b). Very high densities frequently occur in marine shallows; annual average numbers of infusorians in the surface plankton of Denmark (Wadden Sea) range around 10,000–15,000 organisms/liter (Grøntved, 1962). Ivleva found a dense infusorian monoculture of F a b r e a   s a l - i n a   in very brackish waters in large pools isolated from Omega Bay in the Sevastopol area. The high density of this infusorian (I calculated about 90,000 organisms/liter) persisted throughout June 1969. During brief periods of their isolation, tidal puddles in the Concarneau area [Bay of Biscay] contain rapidly growing populations of S t r o m b i d i u m   o c u l a t u m   which reach densities of 200,000–250,000 organisms/liter (Faure-Fremiet, 1948 a).

Tintinnids have long been regarded as the typical and most important group of marine infusorians on the basis of net catches, which evidently retain only such relatively large, shell-bearing infusorians. Works published during the last decade, however, unanimously stress the dominant role among the marine infusorians of small

Oligotricha, especially the genus Strombidium. Tintinnids do not account for more than 5—10% of the total number of infusorians in the marine littoral (Margalef, 1963 b, 1968; Zaika and Averina, 1968). Their number increases slightly away from shore but rarely exceeds 20% of total numbers (Beers and Stewart, 1967; Zaika, 1969 b). The maximal number of infusorians is usually in the subsurface layer, decreasing rather sharply with depth.

Average numbers of infusorians in freshwater appear to be higher than in the marine littoral. In the Rybinsk reservoir, for example, infusorian numbers in the surface range from 77,000 to 276,000 organisms/liter during the summer; along the shore the density of Tintinnidium fluviatile alone can reach 378,000 organisms/liter (Mordukhai-Boltovskaya, 1965). Mass concentrations of infusorians develop even in conditions of Lake Baikal, where maximal density of Marituja pelagica at a depth of 25 m in the Selenga area was found to be 260,000 organisms/liter (Eggert, 1968).

Such cases of exceedingly high densities lead to the important conclusion that for infusorians and other microorganisms with high reproduction potential and extreme limits, the average abundances provide a rather incomplete idea of their role in the community. Reference to average densities alone can lead to underevaluation of the capacity of infusorians to exploit suitable substrates. Abundance of microorganisms increases rapidly near high food concentration, causing an intensive flow of matter and energy through this link of the community. On exhaustion of the food supply, the abundance returns just as rapidly to comparatively low normal levels.

2. REPRODUCTION RATE OF INFUSORIANS

Discussing the role of different groups of organisms in biogeo-chemical processes proceeding in the biosphere, Vernadskii (1965) stressed the great significance of obtaining and analyzing informa-tion on reproduction rates of different species. To his surprise, pertinent information turned out to be meagre. In fact, information on the division rate of many species, including infusorians, had already appeared in the first third of the present century; however, being scattered in many journals, it had not been reviewed and remained largely unknown to researchers interested in the rate of transfer of matter and energy in complex bio- and ecosystems.

As pointed out in discussing methods for determining production of microorganisms, the cell division rate is one of the important parameters in the study of productivity of protozoans and bacteria for evaluating their functional role in the community. Naturally, Vernadskii's idea about the need for collection and systematization of data on the reproduction rate of organisms is dear to ecologists working in the field of productivity.

To collect factual material on the reproduction rate of marine infusorians, we studied several species from the Sevastopol Bay in the Black Sea and two species from the Red Sea (Zaika and Averina, in press). Although experimental conditions did not allow for control of the physicochemical composition of the medium or of food conditions, we tried to imitate natural conditions as closely as possible, so that conclusions can be drawn on the reproduction rates of these infusorian species in the sea on the basis of the results. Prolonged and continuous microscopic observations led to recording of the reproduction rate of the infusorian Uronema marinum for three consecutive generations. It was established that this species divides every 111—121 minutes (average 116 minutes) at 23—25°C in the presence of abundant bacterial food. Of all infusorian species examined so far, Uronema marinum has the highest reproduction rate.

Existing data on maximal division rates of infusorians have been reviewed (Zaika, 1970 a). Table 2 summarizes reproduction rates of infusorians obtained from original studies and all the known literature.

It should be taken into account that data included in Table 2 were frequently obtained during research into topics other than the reproduction rate. Therefore, the authors did not always indicate the food conditions, temperature, etc., at which the reproduction rates were recorded. It is also known that infusorians show definite life cycles in natural conditions with periods of intensive division alternating with mass conjugation or formation of resting cysts.

The reproduction rate of infusorians is a variable quantity, which can decrease to zero in adverse environmental conditions. Therefore, the material presented shows peak reproduction rates of the species during periods of active division in given environmental conditions. Among reproduction rates indicated by the authors, the peak values were selected since they reflect the optimal potentialities of the different species. The species are listed in Table 2. Despite the variety of experimental conditions, many species show comparable maximal reproduction rates of a very large magnitude. The list includes marine and freshwater infusorians, motile and sessile forms and even parasites (Ophyroscolecidae).

Analysis of the data presented in Table 2 shows that about 14% of the infusorians examined within the temperature range of 20—30°C divide less than once every 24 hours, 36% divide 1—2 times,

TABLE 2. Maximal reproduction rates of infusorians (simple division)

Species	°C	Time between successive divisions, hours	Number of divisions per 24 hours	Source
Aspidisca angulata Bock	20	5	4.8	Fenchel, 1968
Blepharisma undulans Stein ..	30	8–12	2–3	Stolte, 1924
Ditto	?	13.2	1.81	Richards, 1929
Chilomonas paramecium	24	7	3.5	Richards, 1941
Ditto	26–30.5	7.1	3.36	Ibid.
Chlamydodon triquetrus O.F.M.	20–22	10	2.4	Pavlovskaya, 1969
Colpidium colpoda (Ehrb.) ...	23–25	7.9	3.1	Hetherington, 1934
Colpoda steini Maupas	30	3	8	Proper, Garver, 1966
C. sp.	19.3	24	1	Adolph, 1929
Ditto	21.6	18.4	1.3	Ibid.
"	26.5	13.3	1.8	"
Condylostoma patulum Clap.et Lachm.	20	46	0.5	Fenchel, 1968
Didinium nasutum (O.F.M.) ...	21	4.9	4.9	Beers, 1929
Diophrys appendiculatum Ehr.	23–25	16	1.5	Zaika, 1970 a
D. scutum Duj.	23–25	13–14	1.8	Ibid.
Ditto	27	12	2	Fenchel, 1968
Epiclintes ambiguus O.F.M. ..	23–25	10	2.4	Zaika, 1970 a
Ditto	20–22	13.4	1.8	Zaika and Pavlovskaya, 1970
Euplotes neapolitanus Wich. .	?	13.7	1.75	Wichterman, 1964
E. patella	?	21	1.1	Cohen, 1934
E. trisulcatus	23–25	6–9	2.7–4	Zaika, 1970 a
E. vannus O.F.M.	23–25	12	2	Ibid.
E. vannus O.F.M.	?	8	3	Borror, 1963
Euplotes sp. 1.	23–25	24	1	Zaika, 1970 a
Euplotes sp. 2.	23–25	12	2	Ibid.
Frontonia marina Fabre-Dom. .	13	34	0.7	Oberthür, 1937
Gastrostyla steinii Engelm. ...	6–8	120	–	Weyer, 1930
Ditto	16	19.5	1.2	Ibid.
"	21	13.7	1.8	"
"	26	11	2/2	"
Glaucoma ficaria Kahl.	?	8	3	Johnson, 1936.
Histrio complanatus Stokes ...	?	7.3	3.3	Richards, 1929
Holosticha diademata (Rees)..	23–25	10	2.4	Zaika, 1970 a
Keronopsis rubra (Ehrb.)	22–26	16–17	1.4–1.5	Zaika and Pavlovskaya, 1970
Ditto	20	32	0.7	Fenchel, 1968
Lacrymaria marina Dragesco ..	15	40	0.6	Ibid.
Litonotus lamella (Ehrb.)	23	12	2	"
Paramecium aurelia Ehrb. ...	21	34	0.7	Woodruff, Baitsell, 1911
Ditto	28	10	2.4	Mitchell, 1929
"	28	13.7	1.7	Woodruff, 1932
"	27	6	4	Whitson, 1914

TABLE 2. Contd.

Species	°C	Time between successive divisions, hours	Number of divisions per 24 hours	Source
Paramecium bursaria (Ehrb.) ...	?	60	0.4	Loefer, 1936
P. calkinsi Woodruff	?	30	0.8	Parker, 1927
P. caudatum Ehrb.	25 (?)	11.4	2.1	Richards, 1941
P. caudatum (Ehrb.)	25–28	13.3	1.8	Richards, 1941
Ditto	26	10.4	2.3	Ibid.
P. lanceolata	?	15	1.6	Greenleaf, 1926
P. multimicronucleatum Pow et Mitchell	27	21.8	1.1	Stranghöner, 1932
Pleurotricha lanceolata (Ehrb.).	?	6	4	Batisell, 1914
Spathidium spathula O.F.M. ...	?	12	2	Woodruff, Spencer, 1924
Ditto	?	10	2.4	Woodruff, Moore, 1924
Stentor coeruleus Ehrb.	18–20	30	0.8	Hetherington, 1932
Ditto	25–28	24	1	Schuberg, 1891
Stylonychia mytilus Ehrb.	?	48	0.5	Machemer, 1964
S. pustulata Ehrb.	?	10	2.4	Parker, 1927
Ditto	?	9.6	2.5	Baitsell, 1912
"	?	8.8	2.7	Greenleaf, 1926
"	25.2	6.4	3.7	Richards, 1941
"	25 (?)	4.8	5	Ibid.
Tetrahymena pyriformis (Ehrb.).	24–25	2.8	8.7	Hetherington, 1936
Ditto	?	2.3	10.3	Scherbaum, Rasch, 1957
Uroleptus mobilis Engelm.	?	14	1.7	Calkins, 1919
Uronema acutum Buddenbrock ...	23–25	3–4	6–8	Zaika, 1970 a
U. marinum Duj.	23–25	1.9	12.6	Ibid.
Ditto	20	2.5	9.6	Fenchel, 1968
Uronychia transfuga O.F.M. ...	23–25	8–10	2.4–3	Zaika, 1970 a
Vorticella nebulifera O.F.M. ..	23–25	13–16	1.7	Ibid.
Zoothamnium altarnans (Clap. et L.)	21	8–12	2–3	Faure-Fremeit,1930
Ophryoscolecidae sp. sp.	39	10–14	1.7–2.4	Westphal, 1934
Oxytricha fallax Stein	?	12	2	Baitsell, 1914

26% 2—3 times, and 24% more than 3 times. In a narrower temperature range (20—25°C), it becomes evident that the small forms generally reproduce more rapidly, e.g., the unicellular algae. Indeed, reproduction rates higher than 3.5 divisions per 24 hours were found mainly in infusorians having volumes no greater than 25×10^{-6} mm^3 (Uronema acutum, U. marinum, Tetrahymena pyriformis, Didinium nasutum and

Stylonychia pustulata). Most other infusorians have
volumes exceeding 70×10^{-6} mm^3. A more detailed analysis of the
relationship between reproduction rate and body volume in infusorians was published by Fenchel (1968).

A comparison of infusorians to unicellular algae shows that their
reproduction rates are not inferior. The reproduction rate of forms
such as U r o n e m a is quite equal to that of bacteria (some species
divide every 20—30 minutes). The reproduction rate of infusorians
has been studied so far in species growing well in laboratory conditions. Hardly anything is known about typical pelagic forms or
about large interstitial infusorians. Nevertheless, the summary in
this section provides a rather clear idea of the productive capacities
of infusorians. The information on the high reproduction rates of
small infusorians, feeding mainly on bacteria, is a convincing argu-
ment for the enormous productive capacity of bacteria (a capacity
often higher than is assumed merely on the basis of existing microbiological data on the reproduction rate and productivity of bacteria).

3. PRODUCTIVITY OF INFUSORIANS
AND BACTERIA

A rather simple method was described in Chapter III for determining specific production of microorganisms multiplying by
binary fission. The value of C can be easily calculated from the
ratio $C = \dfrac{0.693}{g}$ in which only the generation time g must be known.
The diurnal values of C for some values of g are presented below.

g (hours)	1	2	4	8	12	20	40	50	80
C	16.6	8.3	4.1	2.1	1.4	0.83	0.41	0.34	0.22

From the reproduction rates of infusorians given in the last
section, an idea of the productive potential of different species
under favorable conditions is easily obtained. U r o n e m a m a r i -
n u m had the greatest specific production among the species
studied. The value of C for this species was 8.3 at 20—25°C
with abundant bacterial food in the medium. Values of C greater
than 4 are typical for T e t r a h y m e n a p y r i f o r m i s , C o l p o d a
s t e n i n i i and U r o n e m a a c u t u m . About half the infusorians
examined have values of C less than 1. Thus, the high productive
potential of infusorians are realized only in optimal conditions.

In cases where the biomass and reproduction rate are known,
production can be estimated. According to original data, for

example, the infusorian biomass in Sevastopol Bay in summer is 76 mg/m^3 at the neck of the bay and 242.5 mg/m^3 close to shore. Taking into account the species composition, summer temperatures and food regime, the diurnal specific production can be assumed to be equal to 1 (this corresponds to $g = 17$ hours). With $C = 1$, the infusorian production for four summer months (May—August) equals 9.1 g/m^3 in the neck of the bay and 29.1 g/m^3 near the shore. By comparison, the annual production of all copepod species in the neritic zone of the Black Sea equals 1.67 g/m^3 (Greze and Greze, 1969).

Average reproduction rates of infusorians in the open seas cannot be determined accurately. Infusorians living on local detritus concentrations undoubtedly divide at a high rate close to the potential limit for the given temperature. At the same time, in conditions of food deficit, infusorians divide very slowly or not at all (Pavlovskaya, 1969; Zaika and Pavlovskaya, 1970). To some degree, abundance of infusorians reflects conditions and division rate in the given biotope. Since their density in the open sea is fairly low (not exceeding 100 organisms/liter at the surface, and decreasing sharply with depth), the average reproduction rate in such conditions must likewise be low and C must not exceed 0.5 on the average. As a rough estimate for surface water in the ocean, the density can be taken as 50 infusorians/liter, the diurnal value of $C = 0.5$ (i. e., $g = 30$ hours) and average individual weight as 1×10^{-5} mg. Here, production would be 0.25 mg/m^3 per day or about 90 mg/m^3 per year.

In the marine littoral, and in freshwater bodies enriched with organic matter, abundances may reach tens and hundreds of thousands per liter. In such conditions, the division rate is obviously higher than in the open sea. Assuming that $C = 1$, as in the case of Sevastopol Bay, then with an average individual weight of 1×10^{-5} mg we obtain an average annual infusorian biomass of 100—150 mg/m^3 and annual production of 36—55 g/m^3 for surface plankton of the Wadden Sea. Infusorians number 77,000—276,000 organisms//liter in the surface of the Rybinsk reservoir in summer and, consequently, with a biomass of about 1 g/m^3 their diurnal production will be also 1 g/m^3 (30 g/m^3 per month).

In examining the development of the infusorian Z o o t h a m n i u m on planktonic crustaceans in the Black Sea, I found, from data of 3 March 1967, an average density of 16 infusorians on each A c a r t i a   c l a u s i. This corresponds to about 0.5% of the fresh biomass of A. c l a u s i. Assuming that at this time the interval between two successive divisions of Z o o t h a m n i u m   s p. is about 5 days, the specific production of the infusorians would be 0.14.

The following values are typical for A. c l a u s i at this time and place: biomass 2.3 mg/m^3, specific production 0.044 (Greze and Greze, 1969). It is now easily determined that the diurnal production of the crustacean is 0.1 mg/m^3 and 0.014 mg/m^3 for the infusorian. The diurnal production of epibiotic infusorians during periods of massive development corresponds to about 14% of the diurnal production of their crustacean hosts.

A comparison of all materials on infusorians leads to the conclusion that food is the main factor limiting their numbers in the sea. Indeed, data on feeding of planktonic animals in the Mediterranean Sea (Zaika, Pavlova and Kovalev, 1969) show that the predator effect is hardly limiting, at least with respect to tintinnids.

The reproduction rates attest to the fact that at summer temperatures the infusorian concentrations increase rapidly. Finally, data on the relation between the reproduction rate and food concentrations and for ration values established for a series of species dividing at peak rates indicate that with abundant food infusorians reproduce maximally in the sea (at the given temperatures). From the viewpoint of turnover of matter in marine ecosystems, this means that infusorians usually consume all available food (this, naturally, does not involve total destruction of food organisms), thus limiting the numbers of bacteria and small algae.

It should be noted that data on infusorian productivity indirectly provide information on the productivities of other microorganisms, particularly unicellular algae and bacteria. With respect to division rate and specific production, infusorians are not inferior to planktonic algae whose reproduction has been discussed in a number of works (Myers, 1951; Morozova-Vodyanitskaya and Lanskaya, 1959; Lanskaya, 1961; Williams, 1964; Jitts et al., 1964). Consequently, algae and infusorians cannot long exist in the plankton at equal biomasses. In such conditions the infusorian production equals primary production which cannot, however, meet the food requirements of the infusorians since the ratio K_1 between the growth increment and food consumed for infusorians ranges from 0.3 to 0.5. The specific rate of consumption of organic matter by infusorians, therefore, exceeds the specific rate of production of organic matter by algae. The average infusorian biomass must be far lower than the algal biomass (unless still other sources of organic matter are present in the system). This is usually the case. The biomass of infusorians equals 0.08 g/m^3 and that of the phytoplankton is estimated at 1—2 g/m^3 during summer in the neck of Sevastopol Bay (Zaika and Andryushenko, 1969). Diurnal primary production in the bay equals 0.7—1.4 g/m^3 (Finenko, 1965);

consequently, the specific production of algae in these conditions is slightly less than 1 (taking $C = 1$ for infusorians). This means that in summer infusorians in the bay consume an amount of food equal to 10—20% of the primary production of planktonic algae.

A comparison of the division rate of infusorians with available data on the bacterial reproduction rate in water bodies reveals a certain discrepancy. Sorokin (1967), on reviewing data of different authors on the average generation time of bacteria as a whole, concluded that g equals 10—15 hours in eutrophic waters, 20—30 hours in mesotrophic waters and 50—100 hours in oligotrophic waters. These average values for bacteria are practically useless for determining population dynamics of infusorians or their participation in trophic chains in water bodies. On the remains of planktonic animals, the infusorian U r o n e m a divides every 2—3 hours and consumes no less than 1,000 bacteria during its life (Finenko and Zaika, 1968). Bacteria developing on such substrates obviously have higher reproduction rates than the infusorians consuming them.

Many workers have stressed the high rate of division of saprophytic bacteria. Ivanov (1955) found that these bacteria divide every 1.3—5.3 hours in puddles at a fish farm and every 2—4 hours in the shore area of the Rybinsk reservoir (Novozhilova, 1957). Calculating average rates of reproduction for bacteria as a whole, microbiologists emphasize that saprophytic bacteria constitute a negligible fraction of the total population (often less than 1%).

The total production of bacteria can be calculated if their average division rate and total numbers are known. Such generalized values are quite acceptable if all bacterial species are considered as equivalent for the particular purpose of the study (for example, for calculating the food supply of organisms consuming bacteria).

In other cases it is necessary to determine separately the numbers, division rate and production of the functionally different groups of bacteria. This necessity arises when one attempts to determine the rate at which bacteria process a substrate of given biochemical composition, the rate of formation of given bacterial metabolites, or the rate of flow of matter from the substrate through bacteria to their consumers. In such cases the use of averaged values (average division rate, average specific production) can lead to erroneous conclusions.

Chapter V

SPECIFIC PRODUCTION OF POPULATIONS
OF AQUATIC INVERTEBRATES

Many workers have devised methods for determining production
and have applied them to various animals, mainly crustaceans.
These workers have now not only yielded basic schemes for calculating production indexes, but also for estimates of values of production and specific production of many species populations.
Special advances were made on the freshwater crustaceans —
planktonic copepods, cladocerans, benthic amphipods and chironomids — representatives of which groups served as test objects
for various methods of production determination. As a result,
the production indexes of these groups are known accurately and
in detail.

The choice of experimental objects has been determined by the
role of these species in the diet of commercial fishes. However,
values of production and specific production form the actual basis
for establishing the patterns determining the level of productivity
of animal populations. Hence it is necessary to calculate values of
production and specific production for animals representing different
phylogenetic, ecological and/or physiological groups, and to study
thoroughly in different populations the nature of variation in production
and specific production with respect to changes in various populational and environmental factors.

Thus, it is necessary both to expand and deepen research on
production. These two approaches have somewhat different requirements in terms of accuracy of production estimates. The fact is that
the specific productions of species of different taxonomic or
ecological groups may differ considerably. Preliminary establishment of the principles involved requires only a rough estimate of
the production indexes. Exact determinations characterize actual,
particular cases since these relate to given temperatures, actual
conditions of food, growth and reproduction, and specific age structure during the given period. At the same time, it is evident that a
comparison of the productivity, for example, of fishes and infusorians, can only be made using average values characterizing each

group as a whole, or, preferably, the limits of these indexes in different species of each group under different, actual conditions.

The specific production of a single population usually varies within comparatively narrow limits (Zaika and Malovitskaya, 1967; Greze et al., 1968). Accordingly, research for determining the variability of the production of a given population (i. e., productivity research "in depth") requires more accurate procedures for calculating production indexes. In this light, it is easy to understand why some workers criticize calculation methods containing apparently negligible errors and, at the same time, publish obviously rough calculations of production. Before using a production estimate, one must know whether the calculation method was approximate or relatively accurate; both methods may give useful information for solving different problems.

The differences in knowledge on the productivity of various groups of aquatic invertebrates are quite evident in this chapter. The specific productions of some species of poorly known groups were roughly estimated, whereas only the most reliable values of specific production are indicated for well-known crustaceans.

In this chapter, factual data will be presented on the productivity of aquatic invertebrates; the most important comparative index for this is specific production. Therefore, the diurnal value of C of the species will be presented with brief indications of the state of the population during the test period and the method used for calculating specific production. Unless otherwise mentioned, the amounts given represent diurnal values of the specific production and of the specific rate of weight increase. Information is also necessary on the length of the life cycles, when possible, since these values are widely used in Chapter VI. The maximal life span, the age reached by 5—10% of the initial population is noted, whenever known. In most cases, unfortunately, it was impossible to determine the life span in such terms. Therefore, the life span values indicated below only reflect partly the maximum age reached by some part of the animal generation.

1. SPECIFIC PRODUCTION OF ROTIFERS

Galkovskaya (1963, 1968) has done much work on the productivity of many species of freshwater rotifers. The methods of calculation that she used were discussed in Chapter III. It was indicated that with some corrections these methods are quite acceptable. However, the actual calculations of Galkovskaya raise a number of

questions. First, different values of reproduction rates of the same
rotifer species were presented in the two papers on the basis of the
same initial data (Galkovskaya applies the term "reproduction rate"
to the index y_2, which is reciprocal to the doubling time T'). The discrepancies in the values of y_2 in the two publications (1963 and 1968,
respectively) are as follows: A n u r a e o p s i s   f i s s a, 0.85 and 0.33;
E u c h l a n i s   d i l i t a t a, from 0.49 to 0.82 and from 0.25 to 0.27;
E p i p h a n e s   s e n t a, 0.40 and 0.25. It is impossible to say which
values are correct. Second, the quantity y_2 in the equation
$P_t = y_2 \cdot \bar{N} \cdot t$ used by Galkovskaya should correspond to specific
production (see Chapter III). Galkovskaya calculated production
by using y_2, and afterwards determined the " P/B coefficient." With
the latter index it is also possible to obtain y_2. However, for
A s p l a n c h n a   p r i o d o n t a   the value of y_2 indicated is 0.33
(Galkovskaya, 1968), but the P/B for summer months (Lake Naroch')
ranges from 0.11 to 0.26, averaging 0.16. It is impossible to understand why the value of P/B is half that of y_2 for the same species
and temperature. The values of specific production indicated for
all the other rotifer species are also very low, and lower than the
corresponding values of y_2, which itself appears doubtful.

The general impression is that Galkovskaya has made serious
errors in her actual calculations and has obtained values of C which
are too low. It is noteworthy that Bregman (1968) calculated the
specific production of A . p r i o d o n t a in Lake Drivyaty using
the growth curve and age composition of the population, and obtained
$C = 0.21 - 1.32$ (average about 0.5). This is roughly three times
more than that obtained for this species by Galkovskaya. Therefore,
it seems impossible to use the results obtained by Galkovskaya.

A s p l a n c h n a   p r i o d o n t a Gosse, Lake Drivyaty (Belorussia).
Bregman (1968) indicated diurnal values of C which are averages
for 5- to 10-day periods. Observations were made in summer at
water temperatures of 19.8—22.0°C. The oldest age groups usually
predominated in the population. Most values of C ranged from 0.21
to 0.60; $C = 1.32$ was obtained for two periods in August and
September. The average diurnal value of C in summer is 0.5.

B r a c h i o n u s   r u b e n s Ehrbg. (in basins). Vasil'eva (1968)
cultured these rotifers in open basins using algae (C h l o r e l l a,
S c e n e d e s m u s) as food. The result was a highly intensive culture of rotifers. At 21—24°C females live for 4—7 (average, 5) days
and lay 12—27 (average, 15) eggs during this period.

The specific production can be calculated according to the equation $C = \dfrac{0.693}{T'}$. The doubling time of the population will be defined

as the ratio between the life span of the females and the number of their progeny. Considering the two cases of 60% and 100% egg survival and using the initial data given above, this method yields values of $C = 1.26-2.1$.

Vasil'eva (1968) published curves showing the increase in the initial biomass of rotifers in basins with different food regimes and temperatures. From the equation $\frac{dB}{dt} \cdot \frac{1}{B} = \frac{\ln B_2 - \ln B_1}{t_2 - t_1}$ it was possible to calculate the diurnal rate of the biomass increment. This value is 0.4. Since elimination of biomass was not taken into account, the diurnal value of C is obviously greater than 0.4.

Brachionus calyciflorus Pall. (in cultures). Maksimova (1968) presents results of intensive culturing of this rotifer. The biomass increase from 4 to 200 mg/liter in 2—3 days (at 20—24°C) is an example of the very high growth rate. According to my calculations, the specific rate of increase of the initial biomass is no less than 1.3 in such conditions. Consequently, the diurnal value of C exceeds 1.3.

Maksimova (1968) applies the term "production" to the increment in the initial biomass divided by the length of the experiment in days, i. e., $\frac{\Delta B}{\Delta t}$. The corresponding values for different experimental variants are given together with the biomass. Taking $\frac{\Delta B}{\Delta t}$ for the respective values of B, one could calculate C. In different variants of the experiment, C is close to 0.5 during periods of optimal development of the rotifer. C is actually greater because the eliminated biomass was not taken into account; the calculation was made according to the yield of the system in which elimination must be taken into account.

Synchaeta baltica Ehrbg., Sevastopol Bay (Black Sea). This rotifer is common in the winter plankton of the Black Sea. In conditions of Sevastopol Bay peak development is usually reached in January at a water temperature of 10°C. Data for 1952—1953 were supplied by T. S. Petipa for characterizing the numerical development of this species in the plankton of the bay.

Date	Organisms/m³	Date	Organisms/m³
24 November	310	14 January	1,250
8 December	300	9 February	25
26 December	1,278	7 March	66
6 January	34,875		

Early in 1965 I examined plankton samples from the bay to determine the ratio between egg-bearing and nonbearing females. This ratio reflects the reproduction rate because during intensive oviposition the proportion of females with 1, 2 and 3 eggs successively increases. Females with 2 and 3 eggs appeared in the plankton on 29 January, and represented 21% of total population numbers; a single egg was found in 26% and 53% consisted of juveniles and females without eggs. Intensive reproduction took place during the first 10 days of February. Later the rate of reproduction decreased. Females with 2—3 eggs disappeared by 16 February. By 2 March, only 7% had one egg. There was a parallel decrease in total numbers of rotifers in the plankton.

The following data were obtained in February 1969. The percentage of individuals with different numbers of eggs is indicated. At least 100 specimens were examined in each sample.

	12 Feb.	14 Feb.	18 Feb.	19 Feb.	20 Feb.	21 Feb.
Larvae and females						
without eggs	90.3	74.7	60.3	69.0	71.0	73.4
Females with 1 egg ...	6.5	20.0	15.9	28.5	25.5	17.4
Females with 2 eggs ...	3.2	5.0	23.6	1.8	3.5	7.3
Females with 3 eggs ...	0	0.5	0	0.7	0	1.9

The duration of embryogenesis was determined experimentally; it was about 40 hours (1.7 days) at temperatures of 10—13°C. Although incomplete, these data allowed for an estimate of the specific production of rotifers, using equations $y_1 = \dfrac{F}{ND_1}$ and $C = \ln(y_1 + 1)$ (see Chapter III). Since the numbers of females with different numbers of eggs are expressed in percentages, the total number of eggs per 100 specimens (regardless of age) can be calculated. Then, for $D_1 = 1.7$ the calculation of y_1 involves dividing the total number by 170. For January 1965 we obtain $C = 0.36$, and for February 1969 $C = 0.08-0.32$ (average 0.19).

From these data it is evident that under favorable conditions the productivity of rotifers is extremely high. A high specific production can be expected in all small animals comparable, in terms of life span and fecundity, to rotifers, regardless of their taxonomic position. Unfortunately, the productivity of such animals has been poorly studied.

2. SPECIFIC PRODUCTION OF ANNELIDS AND MONOGENETIC TREMATODES

Such groups of worms as nematodes, polychaetes, oligochaetes and turbellarians are abundant in various benthic biotopes and therefore their productivity must be investigated. However, there is practically no information on the weight increase and age structure of worm populations. As far as I know, calculations of the production of worm species have been made only by Borutskii (1939 b), Poddubnaya (1963), Gavrilov (1969), and Richards and Riley (1967).

Data on the growth of some polychaete and oligochaete species were collected and used for a rough calculation of their specific production. Since data on the size structure of the population were lacking or insufficient in all these cases, it was assumed that the specific production of a population cannot exceed the range of variation of the specific rate q_w of individual weight increase. Where a population consists of different age groups it is naturally assumed that the value of C exceeds the minimal values of q_w, but is less than the maximal ones of q_w. Consequently, having evaluated the nature and range of the variation of the specific rate of weight increase during ontogenesis, the possible limits of the variation of C of the population can be determined. Such preliminary estimates provide valuable information on the production capacity of the population.

Ampharete acutifrons (Long Island). Richards and Riley (1967) calculated the production of this species and the value of specific production over the year. The temperature here varies from 0.6 to 21.4°C during the year. The population consists of individuals ranging in length from 2 to 22 mm; medium-sized animals (7—14 mm) were commonest in nearly all samples. There are two reproduction peaks: spring (main) and autumn. C equals 4.58 over the year; consequently, the diurnal average of C for the year is 0.012.

Nereis diversicolor O.F.M. (Denmark coast). Smidt (1951) published a linear growth curve of this species based on material collected in 1947 relating to changes of the average size of specimens in the population. Since the main reproduction peak occurs in late April and early May and is rather brief, the population has a simple size composition; the average individual size increases between June and September with a considerable range of variation. In 1947 the composition of the population changed as follows (size limits in mm):

18 June	22 July	29 Aug.	10 Oct.
0.5–10	3–17	7–42	19–38

From the linear growth curve, I calculated specific rates of weight increase assuming that $q_w = 3\,q_l$. The following values of q_w were obtained within the 3.2—26-mm size range:

Length, mm	3.2–7.2	7.2–21.6	21.6–26
$\bar{q}_w$	0.033	0.09	0.012

Comparison of the range of individual sizes in the population with values of q_w shows that the diurnal specific production ranges in summer from 0.03 to 0.07.

Scoloplos armiger O.F.M. (Denmark coast). Using the same procedure as before, Smidt (1951) obtained the linear growth curve of this species on the basis of material collected in 1946. I obtained the following values of q_w for individuals of different length by the above method:

Length, mm	2–6	6–14	14–18
$\bar{q}_w$	0.09	0.045	0.009

It follows that the diurnal summer value of C ranges from 0.01 to 0.09.

Harmathoë imbricata L. (Barents Sea, Murman). Streltsov (1966) provides a curve of the weight increase of this polychaete. Average annual water temperature in the area observed is 4.3°C (in summer, it reaches 7.5°C). On the basis of the weight increase data, I calculated the value of q_w. The specific rate of weight increase decreases from 0.05 (average q_w for the first year of life) to 0.001 (average q_w for the fourth year of life). The age composition of the population was not examined. According to my estimate, the annual diurnal average for C of the population varies from 0.005 to 0.010.

Limnodrilus newaensis Mich. (Rybinsk reservoir). Poddubnaya (1963) estimated the production of this oligochaete species for three months (mid-June through mid-September) as the sum of the population biomass increment and the biomass of eliminated individuals. However, this calculation relates only to the population of the current year although individuals belonging to the population of the preceding year until August account for more than 70% of the biomass. Therefore it is difficult to determine the total production of the population and specific production on the basis of the calculations of this author.

Since a growth curve of the individuals for the summer is included in the paper, I calculated the specific weight increments q_w:

Weight, mg . to 20	20–60	60–80	80–96
$\bar{q}_w$ 1.1	0.04	0.02	0.004

Considering the proportions of the generations of the preceding and current years, the average diurnal production in summer was estimated at 0.01—0.04.

Enchytraeus albidus Henle (in culture). Ivleva (1953) obtained detailed data on the weight increase of this oligochaete in experimental conditions with abundant food at 18—20°C. Averaging data on the weight increase rate, I found that the specific rate of weight increase decreases with age from 0.25 to 0.01. In natural populations this worm grows less rapidly; according to Ivleva, this species reaches a length of 4.1 cm in experimental conditions compared to 1.2 cm in nature. For this reason the specific production of natural populations of this species is probably closer to the lower limit of q_w.

Oligochaetes (Belorussian lakes, Lake Beloe). Gavrilov (1969) calculated the production of oligochaetes according to biomass loss. In Lake Drivyaty, the summer average of C is 0.038; in Myastro and Batorin lakes, the annual diurnal average of C is 0.014—0.015; in the lakes of the Vitebsk fish hatchery $C = 0.04—0.011$.

Borutskii (1939 b) reports that oligochaetes of Lake Beloe have an annual production $P = 649.6$ kg. The average biomass is 504 kg. This leads to an annual diurnal average value of $C = 0.0035$. A comparison of this value with those mentioned above for different species shows that C is actually probably higher.

SPECIFIC PRODUCTION OF SOME MONOGENETIC TREMATODES

Monogenetic trematodes are parasites of fish. Bykhovskii (1957) discusses their population dynamics with particular reference to the oviparous Dactylogyrus vastator Nebelin and the viviparous Gyrodactylus elegans Nordmann (both species are parasitic on carp). The biological characteristics of these species differ considerably, especially with respect to reproduction and life cycle. With initial data on their reproduction biology (for summer in water bodies in the European part of the USSR), Bykhovskii could calculate the increase in the population in the absence of elimination. Some of the data published in Table 2 of his work are given on the following page.

From these data specific production as $C = \dfrac{\ln N_2 - \ln N_1}{t_2 - t_1}$ can be calculated easily for any 5-day period (see Chapter III). However, the calculation cannot start from the first day because the model does not reflect the complexity of the age composition of the natural

population at that time. For the three 5-day periods from the
15th day onward the following values of C were obtained: D.vasta-
tor 0.28, 0.36, 0.22; C.elegans 0.28, 0.28, 0.28. Thus the
summer diurnal specific production of these trematodes ranges
from 0.2 to 0.3. The maximal life span is 20–25 days for
D.vastator, 13–14 days for C.elegans.

	Numbers	
Time, days	D.vastator	C.elegans
1	1	1
10	6	8
15	31	34
20	130	142
25	760	590
30	2,320	2,452

3. SPECIFIC PRODUCTION OF CRUSTACEANS

With respect to productivity, this is the best known group of
aquatic animals. Most works on the production of aquatic invertebrates deal with planktonic (Cladocera, Copepoda) or benthic
crustaceans (Amphipoda) . I present here for the first time the
values of specific production calculated for species belonging to
different orders.

I shall not indicate all values of specific production published
for crustaceans. For species of the best known orders, only
thorough and reliable results are given, whereas rough or doubtful
estimates are omitted. On the other hand, all the available data
are given for poorly known groups on which accurate information
is lacking.

Daphnia pulex Straus (ponds in the European part of the
USSR). Galkovskaya and Lyakhnovich (1966) made a thorough study
of the productivity of this species in ponds. A growth curve was
obtained in experiments made at 22°C. The summer diurnal average specific production ranges from 0.21 to 0.45. The productivity
of this species in lakes is only $^1/_4$ to $^1/_3$ of this value. This daphnia
lives less than a month.

Daphnia cucullata Sars (Belorussian lakes). Studies were made in lakes having average summer temperatures of 18–20°C. According to Pechen' (1965) the value of C in August is 0.10–0.23. Vinberg et al. (1965) found for five summer months in these lakes (mesotrophic and eutrophic) that the average C is 0.09–0.095. As with the preceding species, these authors calculated the production according to the first variant of the graphic method.

Daphnia longispina O.F.M. (Gor'kii reservoir). Petrova (1967) determined experimentally that at 16.6–22.4°C this crustacean has embryonic development of 2–4 days, postembryonic development of 5–14 days, intervals between ovipositions of 2–4 days and a life span of about 24 days. A growth curve was obtained by calculating growth of the different stages. For all the ice-free months (May–November) growth curves were calculated with respect to water temperature (experimental data on the duration of development were converted to the required temperature using the Krogh curve). Production was calculated according to the first variant of the graphic method (Chapter III). The average diurnal value of C is 0.12 for May–November.

The production of this species was studied in the Ucha reservoir by Lebedeva (1963) who obtained a summer diurnal average of $C = 0.5$. This value seems excessively high, but the brief data prevent verification of the calculation.

Bosmina coregoni Baird (lakes in Belorussia, Gor'kii and Kiev reservoirs). In Belorussian lakes, the average diurnal specific production is 0.14–0.15 in summer (average temperature 18–20°C) (Pechen', 1965). According to Petrova (1967), $C = 0.09$ in May through November in the Gor'kii reservoir. In the Kiev reservoir, Zhdanova (1969) obtained an average diurnal value of $C = 0.10–0.15$ for five summer months, and found that this species has a life span of 27–36 days in laboratory conditions. Pechen' (1965) found the following indexes at 17°C: embryonic period, 2.7 days; from hatching to sexual maturity, 5.9 days; from hatching to the appearance of the new generation, 8.6 days; interval between oviposition, 2.7 days.

Bosmina longirostris O.F.M. (Kiev reservoir). Zhdanova (1969) studied the growth of this species in experimental conditions (the data indicate a water temperature of 11–19°C, but the description of the growth experiments does not mention the temperature). The following results were obtained (in days): embryonic development, 2–3; period of sexual maturation, 3–4; interval between ovipositions, 2–3; life span, 20–25. This species appears in the plankton in March–April and reaches peak density in May. It disappears from the plankton during the first half of July and reappears in late August. It reaches peak density at a

temperature of 17–19°C. Production is calculated by the first
variant of the graphic method. For five summer months C averages 0.14–0.15. The value of C in May is 0.22.

Chydorus sphaericus O.F.M. (Belorussian lakes,
Gor'kii reservoir). According to Pechen', this species has the
following growth and development characteristics at 17°C: embryonic period, 2.3 days; from hatching to sexual maturity, 5.8 days;
from hatching to the appearance of the new generation, 8.1 days;
period between ovipositions, 2.4 days. The summer average
diurnal specific production is 0.13–0.20 (in the Belorussian lakes).

Petrova (1967) found that this species has a life span of about
22 days at 18–23°C. In Gork'ii reservoir this species reaches
maximal density and biomass in August. The maximal diurnal
specific production of 0.33 is obtained in September. The average
value of C for seven summer months (May–November) is 0.19.

Moina rectirostris Leydig (purification ponds, Belorussia).
Kryuchkova (1967) examined the growth of this species at 19–21°C.
The average fecundity of the female is 42.6 eggs. The life span
reached 20 days in the experimental conditions. The specific production was calculated for populations of purification ponds, but the
date of the collection is not indicated. Apparently the material was
collected in summer at a water temperature of about 20°C. The
average diurnal specific production is 0.25.

Diaphanosoma brachyurum (Belorussian lakes).
Pechen' (1965) calculated the diurnal specific production from data
from a number of lakes in August (water temperatures of 17–23°C),
$C = 0.13–0.21$.

Ceriodaphnia reticulata (Jur.) (Belorussian lakes).
Pechen' (1965) gives the following growth and development parameters for 16°C: embryonic period, 4 days; from hatching to sexual
maturity, 9 days; from hatching to the appearance of the new generation, 13 days; period between ovipositions, 4 days. The adult
crustaceans are 0.3 mm long and weigh 0.014 mg. The specific
production in August at 17–23°C is 0.08.

Penilia avirostris Dana (Black Sea). According to
Pavlova (1959) the most intensive development of this species in
Sevastopol Bay is in August. Maximal numbers (more than
2,000 organisms/m³) are attained in late August and early
September at water temperatures of 19–22°C. At the end of September the density of this crustacean in the plankton decreases
sharply (to 20 organisms/m³); in October, P. avirostris disappears from the plankton. During the peak development, young
forms (0.49–0.56 mm) constitute 40–60% of the population. The
parthenogenetic females grow to a size of 0.91–0.98 mm and produce an average of 8 eggs during this period. By autumn the

number of eggs in the brood pouch decreases to 3—4.   Thus,
P . a v i r o s t r i s   develops in the plankton only during 3—4 months
per year; resting eggs are found during the remaining time.   The
maximal life span is about 20—25 days.

According to Greze (1967 b), the specific production of this
species is C = 0.188.   This calculation was made after the second
variant of the graphic method (see Chapter III).

C y c l o p s sp. (Lake Drivyaty, Belorussia).   Pechen' (in
Vinberg, 1968 a, section 5, 2, 1) calculated the specific production of
this species from materials collected in August (water temperature
of 18°C).   The duration of embryonic development is 2.6 days, the
nauplius stage 10.2 days, and the copepodid stage 12.8 days.   This
form lives about a month.   The average diurnal value of C is 0.12
in August.

C y c l o p s k o l e n s i s Lill. (Lake Baikal).   The annual average
diurnal specific production is 0.023 (Greze, 1967 a).   The average
annual temperature in the zone inhabited by this form is 4°C.   In
these conditions, the nauplius and copepodid stages last very long,
about 160 days.   Females produce an average of 400 eggs.   The
numbers of the sexes are equal.   The maximal life span is about
10 months.   The annual average density is 750 nauplii, 1,200 copepodids and 240 adult crustaceans per m^3.

These data were used for calculating specific production by the
second variant of the graphic method.

E p i s c h u r a b a i c a l e n s i s Sars (Lake Baikal).   The annual
average diurnal specific production is 0.027 according to Greze
(1967 a), who made a number of assumptions in plotting curves of
weight increase and individual production.

According to Afanas'eva (1968), the winter generation develops at
an average temperature of 2.61°C and the summer generation at
4.97°C.   In such conditions nauplii develop within 90 days, the
copepodid stage lasts 90 days and adults live 180 days.   The egg
sac of the female contains an average of 22 eggs; the interval
between ovipositions is 20 days and embryonic development lasts
20 days.   The maximal life span is about 7 months.   Using different
methods, Afanas'eva (1968) arrived at an annual average diurnal
value of C = 0.022—0.031.

M o r m o n i l l a m i n o r Giesbr., H a l o p t i l u s l o n g i c r o n i s
Claus (Mediterranean Sea).   Greze (1967 b) calculated the specific
production of these species by plotting the individual growth curve
according to peak displacement on age structure graphs.   These
crustaceans inhabit depths of 300—500 m at 13—17°C and have a life
span of over a year.   Other data are not reported.   The diurnal
specific production is 0.004—0.005.

Calanus finmarchicus Gunner (Barents Sea, Eastern
Murman). Kamshilov (1958) calculated the production of this
species according to biomass loss. This species lives at water
temperatures not exceeding 5–8°C. Mass reproduction takes place
in spring, and density reaches a peak in mid-May. The average
individual weight later increases because of growth, and abundance
declines due to elimination. This simple structure of the population
makes it possible to determine annual production from the biomass
loss. The maximal life span is about 10–12 months. The annual
average diurnal specific production is 0.019.

Calanus helgolandicus Claus (Black Sea). According to
Petipa (1967), this species lives in the Black Sea at depths ranging
from 5–20 to 60–150 m and performs pronounced vertical diurnal
migrations. Water temperatures in the habitat are 6–15°C.
Growth is near-parabolic. Petipa calculated the specific production of this species in summer as the ratio between the total daily
increment of the population and the biomass, expressing all values
in calories, $C = 0.14$.

Limnocalanus johanseni (a lake in Alaska). According
to Comita (1956), these crustaceans live half a year and die by the
winter. Only eggs overwinter. Hatching occurs during the thaw
(in July). The females mature within 16 days. Maximal water
temperature is 10.5°C (in August). Reproduction begins in the
first half of September and continues for 1.5 months.

Greze (1967 a) calculated the specific production of this species
as an average for 2.5 months, finding $C = 0.03$. Using data
published by Comita (1956), I drew a curve of weight increase for
this crustacean. Owing to the simple age structure of the population, production can be easily calculated from the biomass loss on
the basis of the growth and survival curves. For the period of
5.5 months, the life span of this crustacean, a production of
1.930 mg/m³ (fresh weight) with an average biomass of 600 mg/m³
was obtained. The average diurnal specific production for this
period is 0.019.

Centropages kroyeri Giesbr. (Black Sea, neritic zone).
A warmwater species, it develops in the plankton from May to
October. The peak density is in August. The water temperature
is 15–23°C during the period of planktonic life of this species. The
nauplius period lasts 10 days, the copepodid period 27 days, and
the adult stage 58 days. These stages have, respectively, the
following average weights: 0.0006, 0.008 and 0.05 mg. Greze et al.
(1968) determined the seasonal and annual changes of the diurnal
specific production on the basis of materials collected in the
Sevastopol area over several years. The average diurnal specific

production in summer (average temperature 21°C) during different
years lies within the range of 0.09–0.115.

Paracalanus parvus (Claus) (Black Sea, neritic zone).
This species is found the year round, occurring to a depth of
150–175 m, but preferring the surface. Peak reproduction is in
early spring and the beginning of summer. Analysis of data for a
number of years shows that the average diurnal specific production
in summer is 0.08–0.09 (Greze et al., 1968).

Pseudocalanus elongatus (Boeck) (Black Sea, neritic
zone). A coldwater species living at temperatures of 6–15°C and
performing diurnal vertical migrations. It reproduces throughout
the year. Greze et al. (1968) found that $C = 0.14–0.17$ as the
summer average for a number of years.

Oithona minuta (Kricz.) (Black Sea, neritic zone). Occurring the year round, these small crustaceans (adult length 0.4–
0.7 mm) are an important food item for larvae of pelagic fishes.
Greze et al. (1968) calculated that the summer average diurnal
value of $C = 0.11–0.14$.

Oithona similis Claus (Black Sea, neritic zone). A coldwater species, it has $C = 0.07–0.13$ in summer (Greze et al., 1968).

Acartia clausi Giesbr. (Black Sea, neritic zone). This
species reproduces the year round in the Black Sea. Greze and
Baldina (1964) distinguished seven generations during the year.
Reproduction is most intensive during the spring and summer
(May–June). The August generation develops in about 30 days (at
a water temperature of 23.5°C). Experimental data on development
of this species obtained at 20°C were used for plotting curves of
weight increase in different seasons (temperatures were corrected
according to the Krogh curve). The female produces 15 batches of
16 eggs each. The hydrological summer (lasting 127 days) has an
average temperature of 21°C. The average numbers (per m^3) for
the different stages at this period in 1960–1961 were 730 naplii,
406 copepodids, and 173 adults. According to Greze et al. (1968),
the perennial fluctuations of the specific production in summer lie
between 0.12 and 0.17. Petipa (1967) obtained $C = 0.22$ for the
western halistasis of the sea (June).

Copepods of the Sea of Azov. L. M. Malovitskaya calculated
production characteristics for the main copepod species of the Sea
of Azov. Only part of her data was published with respect to the
analysis of the variability of specific production (Zaika and Malovitskaya, 1967). These calculations show the following average
summer values of Centropages kroyeri 0.09–0.15,
Calanipeda aquae-dulcis 0.10–0.20, Acartia clausi
0.05–0.08.

However, the same error was committed in preparing growth curves for copepods of the Black Sea and Sea of Azov — the definitive weight was plotted against the middle (and not the beginning) of the time interval corresponding to the life span of adult crustaceans. With respect to Black Sea species, the correction of this error yielded C values roughly twice the initial ones: A. Clausi 0.044, C. ponticus 0.077 (summer data, Greze and Baldina, 1964). The corrected values of C are, respectively, 0.12−0.17 and 0.09−0.11 (summer, Greze et al., 1968). It can be assumed that C values for the Sea of Azov copepods are also about twice those indicated, but we do not have the initial data necessary for the recalculation.

Here one must note the following: in Chapter VI in the analysis of the relation between specific production of the population and average individual weight, results of calculations of specific production of Sea of Azov copepods are used. As noted above, these values are too low. If all other conditions are equal, however, specific production is directly proportional to the growth rate. Therefore, the necessary recalculation would only shift all the points upward in the graphs without altering their mutual relationship. Consequently, the conclusion reached on the quantitative relationship between specific production and average individual weight remains correct. Moreover, data on many other invertebrate species are also used in the discussion of this topic.

Harpacticoids of the Black Sea. Considering the importance of harpacticoids in different benthic communities, I tried to determine the specific production of species living on growths of Cystoseira barbata Ag. Apart from worms, the population on the Cystoseira includes numerous harpacticoids at numbers reaching 200,000 per kg fresh algal weight. A total of 13 harpacticoid species have been found here, but it is 2−3 species that usually account for 70−80% of the population. Harpacticoids feed mainly on diatoms ("microovergrowth").

Griga (1960) described the development of the Black Sea harpacticoids Nitocra spinipes (Boeck) and Tisbe furcata (Baird) in laboratory conditions. She mainly studied the morphology of different development stages, but indicated with this the times of appearance of different stages and their sizes, making it possible to plot growth curves up to the first copepodid stage. The development of T. furcata was examined earlier by Johnson and Olson (1948) whose data agree with those obtained by Griga (1960) at a temperature of 21−24°C. The animals were fed on dried ground macrophytes.

To obtain more detailed information on the linear growth of harpacticoids, I reared two species (T. furcata and Dactylopodia sp.) at 19−20°C. Diatoms were used as food. The growing

crustaceans were caught and measured at regular intervals. For
constructing growth curves, the sizes of the largest animals of
each batch were used, since age was recorded from the moment of
hatching of the first nauplii in each experiment. Egg-bearing
females of T . fu rc a ta were obtained at the end of the experiments.
The linear growth curves of harpacticoids based on my data and
those of Griga are shown in Figure 8. Griga also obtained data on
the growth of N i t o c r a s p i n i p e s , whose growth is roughly
similar to that found by us in D a c t y l o p o d i a sp.

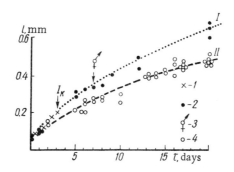

FIGURE 8. Linear growth of Black Sea harpacticoids:

I) Tisbe furcata (1 – after Griga, 1960; 2 – original data, ages of
formation of the first copepodid stage ($l_κ$) and sexual maturation
indicated (3) after R.E.Griga, 1960); II) D a c t y l o p o d i a sp., original
data; 4 – points of experimental results.

From information on the linear growth of these three harpacticoid
species a general, rough estimate of the productivity of harpacticoids
was made. Using the equation $q_l = \frac{\ln l_2 - \ln l_1}{\Delta t}$ specific rates of linear
growth were calculated for the main developmental stages; from the
equation $q_w = 3q_l$ I then obtained the specific rates of weight increase
q_w . Combining material of the examined species, results are as
follows: nauplii, $q_w = 0.85-1.0$; copepodids, $q_w = 0.4-0.9$; adults,
$q_w = 0.10-0.15$.
    The species D a c t y l o p o d i a t h i s b o i d e s , H e t e r o l a -
o p h o n t e s t r o m i and H a r p a c t i c u s g r a c i l i s dominate on
the C y s t o s e i r a in summer (Makkaveeva, 1961). I believe that
the q_w ranges indicated above characterize somewhat the entire
mixed population of harpacticoids on the C y s t o s e i r a . To determine the age structure of the mixed population the proportions of
the different stages in the two samples (13 July 1967 and 5 June 1969)

taken in the Sevastopol area were established, as follows: nauplii, 12−22%; copepodids, 46−52%; adults, 32−36%. A comparison of the q_w values with these data on age structure leads to the value $C = 0.2-0.4$ as a rough estimate of the specific production of harpacticoids (due to individual growth) on Cystoseira communities during the summer.

Gammarus lacustris Sars (Lake Sevan, lakes of the Baikal region). According to Markosyan (1948) gammarids of Lake Sevan populate mainly the littoral zone to a depth of 15 m. Here water temperature reaches a peak in July−August not exceeding 20°C. Gammarus lacustris reproduces throughout its distribution range. Most of the mature females are 9−10 mm long. Oviposition occurs from April through October. At the end of May, 88% of the mature females bear eggs. The young begin to be released from the brood sacs during the first half of June. The number of eggs varies from 2 to 19. Most of the gammarids produce 4 generations during all the reproduction period. Young specimens steadily predominate in the population. On the average, mature individuals account for 33.6% of the entire population over the year. The biomass varies considerably during the year, with the maximum in April and minimum in July. The maximal life span is 2 years. Markosyan (1948) also examined growth and calculated annual production, finding that the annual average diurnal specific production is $C = 0.0055$.

Bekman (1954) thoroughly investigated the biology of this species from materials taken from several lakes in the Baikal area. In Lake Staroe (Angara floodplain), the water temperature rises to 20−22°C in summer. Oviposition begins here during the second half of April. Egg development takes a month. Females deposit the second batch of eggs in June, and most of them die in August−September. Thus, the life span does not exceed 14−15 months (2-year old specimens are very rare). The annual average diurnal specific production of this species in Lake Staroe is 0.0082.

Gammarus locusta L. (Black Sea). Greze and Greze (1969) calculated the specific production of this species. The total weight of the eggs produced by a single female corresponds to 170% of individual adult body weight (assumed to be 50 mg). The winter average diurnal specific production is 0.017, the summer average $C = 0.048$.

Gammarellus carinatus (Rathke) (Black Sea). The specific production was calculated by Greze and Greze (1969). Immature specimens appear close to shore in November at a water temperature of 14−15°C. The first egg-bearing females appear in December at a temperature of 12°C. Reproduction reaches a peak in January−February. The maximum number of eggs found in a

single female was 134. The incubation period at 9—10°C lasts
40—45 days. Males mature in the 7th—8th month and females one or
two months later. The maximum weight of an egg-bearing female is
103 mg. Most of these crustaceans are 2—4 mm long in February—
March (at the age of 2—4 weeks). In May at 12—14°C the
crustaceans migrate to deep waters. At this time, 70% of the
population are 2—3 months old. From May to November, the
crustaceans keep to depths of 20—50 m (temperature usually 7—8°C).
The life span is slightly more than a year. The annual average
diurnal specific production is $C = 0.008$.

D e x a m i n e s p i n o s a (Mont.) (Black Sea). The specific production was calculated by Greze and Greze (1969). Reproduction
continues the year round. Egg-bearing females are most numerous
in spring (March—April) and autumn (September—October). In
March—April, the population consists mostly of older individuals
(5—9 mm long). After spring reproduction, the older age groups
begin to die and disappear from the population by early summer.
Specimens hatching in March—April reproduce in summer. The
maximum number of eggs per female is 81. The first molt occurs
2—4 days after hatching, followed by further molts at intervals of
4—5 days in summer and 8—9 days in winter until the specimens
mature. The mature animals molt every 7—9 days. The average
linear increment between molts is 0.2 mm. Egg-bearing females
have a length of 7—8 mm and weigh 13—16 mg. The life span does
not exceed a year. The diurnal average specific production (C)
equals 0.020 in summer and 0.017 in winter.

A m p h i t h o e v a i l l a n t i Lucas (Black Sea). According to
Greze and Greze (1969), the diurnal specific production of this
species is 0.024 in winter and 0.050 in summer. The maximal life
span is about 10 months.

P o n t o p o r e i a a f f i n i s Lindstr. (lower reaches of the Yenisei,
Kara Sea, Lake Krasnoe, Karelian Isthmus). Greze (1951) analyzed
the productivity of this crustacean in the Yenisei where its life span
is 14 months, and in Lake Taimyrskoe where it survives for
27 months owing to the lower temperature. The summer temperature in the Yenisei does not exceed 20.5°C. The populations in both
cases have the simplest age structure since there is only one brief
reproduction period (in spring). In these conditions Greze successfully applied the method for calculating production by biomass loss.
The values of C are 0.0094 for the Yenisei population (Dudinka area)
and 0.0052 for the Lake Taimyrskoe population. For the Kara Sea,
Greze (1967b) reports $C = 0.01$; this seems to be a printing error,
and the correct value is probably $C = 0.001$.

Kuz'menko (1969) calculated the production of this species in
Lake Krasnoe. Water temperature here is higher than in the lower

reaches of the Yenisei; the annual average is 7.5—7.7°C. The animals grow more rapidly, mature sooner and show a high fecundity, $C = 0.010-0.012$.

Micruropus kluki (Lake Baikal). Bekman (1959) calculated the specific production of this species. The crustaceans inhabit coastal sands of Maloye More. Masses of young forms appear in June—August. The parental generation disappears by winter. The maximal life span is slightly more than a year, $C = 0.0068$.

Micruropus possolskii Sow. (Lake Baikal). Bekman (1962) points out that this species is a warmwater stenobiont found in sunlit eutrophic and mesotrophic water bodies linked to Lake Baikal. The average water temperature for the warmest month is 18—20°C in these waters. This species forms concentrations reaching densities of thousands of specimens per m^2 near shore in the spring. Egg-bearing females appear while the water is still under ice; mass oviposition takes place in May. Reproduction is largely completed by August, after which the parental generation disappears. The maximal life span is about one year. The average annual value of $C = 0.010$.

Gleminoides fasciatus Stebb. (Lake Baikal). Bekman (1962) calculated the production of this species for the population of the open waters of Baikal and the thoroughly sunlit Posol'skii Bay. In the open lake, egg-bearing females are still absent in early June. Progeny appears only in late July from eggs deposited here in June; specimens mature after two growth seasons because the average temperature of the warmest month is 12.4°C, $C = 0.0044$.

In Posol'skii Bay, the life cycle resembles that of the preceding species. During spring and summer, this species forms concentrations near shore at a density often exceeding 10,000—20,000 specimens/m^2. The rapid development at the higher temperature is reflected in the value of the specific production: $C = 0.0080$. The life span is slightly more than one year here.

Hyalella azteca (Sugarloaf Lake, Michigan). On the basis of weekly observations, Cooper (1965) determined the abundance of different stages in the population of this crustacean. Knowing the duration of each stage, he calculated the expected abundance of each group by the next observation date. Since the actual figure is lower than expected because of elimination, mortality would equal this difference. According to Cooper, the biomass of eliminated individuals equals production. Diurnal specific production during the observation period was calculated as $C = 0.032$.

Corophium nobile Sars (Caspian Sea). Osadchikh and Yablonskaya (1968) note that this species produces one generation

per year in the northern Caspian Sea. After a brief reproduction
period, the adults die in May. Thus, the May density peak is due
to the young forms whose further growth is accompanied by a
decrease in density. The population is of the simplest structure.
The life span does not exceed a year. The annual average of the
diurnal specific production is 0.080.

Byblis veleronis Bernard (Pacific Ocean, Washington).
Lie (1968) gives a linear growth curve based on shifting
peaks of age structure graphs. The relationship between body
length and weight is also indicated. The age structure of the
population is given for different seasons. From these initial data
I prepared a weight increase curve from which C was calculated
for two periods (with reference to the age structure indicated for
these periods). Data for January–February and April–May were
used. I found that $C = 0.011–0.012$.

Orchestia bottae M.-Edw. (Black Sea). Biological studies
and production calculations of this species were published by
Sushchenya (1967). In the Sevastopol area, this species live on
macrophytes cast ashore. The average monthly temperature varied
from 18 to 25.3°C from May through October. In December–
February temperatures averaged 1.9–2.7°C. The growth at 20°C
was examined. Growth curves for other temperatures were extra-
polated using the Krogh curve. The reproduction period covers
more than four months, during which females produce 7–8 batches.
The incubation period lasts 12 days at 20°C. The specific production is maximal during periods of mass reproduction and rapid
growth. For July–September, C averages 0.053. The annual
diurnal average of $C = 0.023$.

Acanthogammarus grewingki (Dyb) (Lake Baikal).
Bazikalova (1954) described some aspects of the biology of this
species which is one of the commonest deepwater amphipods of
Baikal. Its density is greatest at a depth of 250–300 m where
water temperature remains 3–4°C the year round. Specimens of
86-mm length weigh 8.5 g (this is the largest Baikal amphipod).
The maximal life span is about 10 years. Egg-bearing females are
encountered throughout the year but are most numerous in November.
Oviposition takes place mainly in October–November. Hatching
is apparently completed by late June. Bazikalova indicated the
size structure of young forms for different seasons. These curves
have distinct peaks which shift toward the greater dimensions with
time. From the peak displacement, some parameters of the
Bertalanffy growth equation were found, using the method described
in my work on the productivity of appendicularians and sagittas in
the Black Sea (Zaika, 1969 a). This method can be outlined as

follows. In the case of S-type growth, linear growth can be expressed, according to Bertalanffy, by the equation

$$l_\tau = l_\infty (1 - e^{-\alpha t}), \tag{60}$$

where l_τ is the individual length at age τ, l_∞ the theoretical limit
of growth (at age $\tau \to \infty$), and α an index remaining constant for
individuals of the populations if growth conditions do not change.
For the time period Δt, during which individual length changes from
l_1 to l_2, equation (60) yields the expression

$$\alpha = -\frac{1}{\Delta t} \ln \left(\frac{l_\infty - l_2}{l_\infty - l_1} \right), \tag{61}$$

from which the value of α can be found. If l_∞ is known, the growth
curve can be completely reconstructed on the basis of fragmentary
data for different growth stages.*
    Analysis of the peak displacement characterizing the size struc-
ture of the population in a series of consecutive samples yields
several pairs of l_1 and l_2 values, each of which can be used for calculating α in equation (61). After these calculations, particular
attention should be devoted to the scatter of the obtained values of α.
The smaller the scatter, the greater the confidence in the result.
The arithmetic mean of α obtained is used for plotting the growth
curve after equation (60).

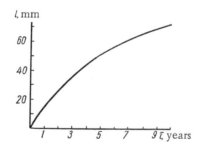

FIGURE 9. Linear growth of the Baikal amphipod
A c a n t h o g a m m a r u s g r e w i n g k i , drawn after
equation (60). Explanations in the text

---

* A similar formula is given in the handbook "Methods for the Determination of the Production ..."
  (Vinberg, 1968 a, p.52), although for weight increase rather than linear growth as in equation (61).
  For this formula,the value of one more parameter must be known.

Analyzing the data of Bazikalova, several values of α ranging from 0.0003 to 0.0007 were calculated. The average value (α = 0.0005) was used for drawing the growth curve. The accepted value of l_∞ is 87 mm. The linear growth curve based on these parameters (Figure 9) agrees with growth data given by Bazikalova. Our materials on the relationship between length and weight yield a curve of weight increase. Using data on the size structure of the population in July, $C = 0.0004$ was obtained as an average for this month.

Asellus aquaticus L. (lakes in Belorussia). From the data of Arabina (1968), the annual average of diurnal specific production of this species is 0.012.

Pontogammarus robustoides (Grimm.) (lower reaches of the Don). Ioffe and Maksimova (1968) examined data on the change in age structure of the population between spring and fall. From their data, I obtained a linear growth curve for specimens of the first (May) generation, and information on the relationship between length and weight served in preparing a curve of weight increase. In tabulated form, the growth data appears thus:

Age, days	20	30	40	50	60	70	80	120	150
Weight, mg	0.4	2.1	8	15	27	38	46	71	85

The maximal life span is 14 months. On 15 May the population consisted entirely of adult specimens of 14–19-mm length, corresponding to $C = 0.005-0.007$. On 9 June, specimens ranging in length from 2 to 4 mm predominated in the population, but one-year-old specimens were rather numerous. In the latter case $C = 0.008$. By 8 July the one-year-old specimens had disappeared; at this time $C = 0.056$. $C = 0.03$ for the age structure recorded on 8 September. The annual average of dirunal specific production is 0.018–0.020.

Balanus balanoides L. (Port Erin, Isle of Man). Moore (1934) studied the growth of this species in units of volume. There are also available data on the age structure of the population. The following three areas are compared: Port Erin (54°N), St. Malo (49°N), and Herdla (60°N). The growth rate in the Port Erin area is intermediate, and the highest value was found in the north, in the Herdla area.

If the specific weight of the individual changes little with age, the specific production can be calculated from the volume which is considered proportional to weight. My calculations are based on the assumption that the specific weight remains relatively constant.

The specific production was calculated for specimens in the uppermost part of the inhabited zone and for the age structure observed on 2 June 1932 (individuals more than 7 mm long accounted for 73% of the total population). The size categories

taken for calculation were larger than the original ones.
$C = 0.0022$ was obtained.

Moore does not indicate the life span, although it is not less than
4 years. Kuznetsov (1960) indicates that this species lives up to
13—14 years in Murman, 10—13 years in the White Sea (although
specimens aged 18—25 years are also found), 7—13 years in Belgium,
and only 2 years in Norway. In the present analysis of specific
production and maximal life span (Chapter VI) a life span of about
8 years was assumed in the area indicated.

Shrimps, Pandalidae (Pacific Ocean, area of Vancouver). Butler
(1964) examined linear growth curves, length/weight ratios and
the size structures of several populations of shrimp species. The
growth curves were plotted from the age of five months onward,
and the younger specimens were disregarded also in the study of
the age structure (they were absent from catches). Many histograms
of age structure from several parts of the examined area and for
different seasons were presented.

The 24-hour specific production of six shrimp species was calculated after data obtained in November 1960. Butler examined
size structure in 0.2-mm interval classes. My own calculations
are based on larger intervals. The females were weighed together
with the eggs, which means that production resulting from reproduction was taken into account. Actually, the values obtained are
probably a little lower than the actual ones because only specimens
older than 5 months were taken into account.

Pandalopsis dispar Rathbun (English Bay). The life span
is slightly more than 3 years, $C = 0.0020$. Pandalus jordani
Rathbun (Vancouver Island). Life span of 3 years, $C = 0.0013$.
Pandalus borealis Kroyer (Indian Arm). Life span 3 years,
$C = 0.0010$. P. platyceros Brandt (Vancouver Island). Life span
slightly more than 4 years, $C = 0.0017$. P. hyspinotus Brandt (Van-
couver). Life span 3 years, $C = 0.0014$. P. danae Stimpson (Buzzard
Inlet). Life span 2.5 years, $C = 0.0020$.

Neomysis americana Smith (Long Island). Production
was calculated by Richards and Riley (1967). The water temperature in the area examined is 14—21°C in June—September, 5°C in
April, 10°C in May. This species is very abundant between late
autumn and early summer, but is hardly encountered during the
summer and early autumn. The body length is 4.0—8.9 mm
(average 6.5 mm) in spring, 3.5—10.4 mm (average 7.0 mm) in
autumn, and 3.5—12.7 mm (average 8.5 mm) during the winter and
early spring. It reproduces the year round. The annual average of
the diurnal specific production is 0.010.

Crangon septemspinosa (Long Island). This species was
studied by Richards and Riley (1967). By the end of the first year,
specimens reach an average length of 12 mm, and 24 mm at two
years. Three-year old specimens are rare. Egg-bearing females

occur almost throughout the year, except for January and November. The yearlings reproduce in August through October, and older animals until the end of the year. The annual average of the diurnal specific production is 0.010.

Pinnixa schmitti Rathbun (Pacific Ocean, Washington). Lie (1968) gives a linear growth curve based on the peak displacement on graphs showing the change of the size structure of the population. The individual length/weight ratio is also indicated. From these initial data, I calculated the value of C for the two months having the greatest difference in size structure of the population. Thus, C equaled 0.0034 and 0.0036 for January and August 1963, respectively.

4. SPECIFIC PRODUCTION OF AQUATIC LARVAE OF INSECTS

Insect larvae often play an important role in the benthos of rivers and lakes. Therefore, many works deal with their productivity. The calculation of the production of heterotrophic animals has special characteristics since the emergence of the imago must be taken into account.

I have not calculated the specific production of insect larvae. All the values given below are taken from Soviet publications. They characterize the productivity of chironomids and chaoborines in water bodies located in the European part of the USSR.

Species	c	Source
Chironomus plumosus L.	0.007	Borutskii, 1939 a, c
" " "	0.007–0.009	Kirpichenko, 1940
" " "	0.007–0.008	Yablonskaya, 1968
" " "	0.019–0.024	Sokolova, 1968
Chironomus anthracinus Zett.	0.008–0.012	Sokolova, 1968
Procladius sp.	0.05–0.07	Sokolova, 1968
Tanytarsus sp.	0.09–0.10	Sokolova, 1968
Polypedilum nubeculosum Meig. ..	0.017–0.027	Sokolova, 1968
Polypedilum bicrenatum Schrnk. ..	0.046–0.048	Sokolova, 1968
Microtendipes pedellus Deg.	0.03–0.04	Sokolova, 1968
Limnochironomus pulsus Walk. ...	0.044–0.07	Sokolova, 1968
Psitolanypus imicola Kieff.	0.013–0.016	Sokolova, 1968
Lauterborniella brachylabis Ed. ..	0.03–0.04	Sokolova, 1968
Einfeldia gr. carbonaria	0.037	Gavrilov, 1969
Chaoborus crystallinus	0.033	Gavrilov, 1969

5. SPECIFIC PRODUCTION OF MOLLUSKS

In recent years several works have been published dealing with production of mollusks (Negus, 1966; Greze, 1967 b; Arabina, 1968; Osadchikh and Yablonskaya, 1968). Information on the production of mollusks is also found in some earlier publications (Kuznetsov, 1948 a, b; Vorob'ev, 1949). Using published data on the growth and age composition of mollusk populations, I have calculated the specific production of some additional species (Zaika, 1970 b; Zaika, 1970 c).

As in preceding sections, productivity is expressed by the average diurnal value of C. Here C represents the annual average. In those cases where the annual mollusk production was already determined by some method, it was divided by the annual average of biomass and by 365 (regardless of the number of months during which the mollusks grow). In other cases the following algorithm was adopted for calculating C.

1. From data on the weight increase of mollusks, the average specific rate of weight increase was calculated for each year of life (at age i), i.e.,

$$q_i = \frac{2.3\,(\log w_{\tau+1} - \log w_\tau)}{365} \tag{62}$$

where w_τ is the weight of the mollusk at the age of τ years, $w_{\tau+1}$ the weight at the age $\tau + 1$ years. The value of q was also calculated for shorter time intervals for mollusks living less than a year.

2. Knowing the age structure of the population at a given moment or period, the production rate (i.e., diurnal production P) was calculated thus:

$$P = \Sigma q_i B_i, \tag{63}$$

where B_i is the biomass of the i-th age group.

3. The next step was calculation of C:

$$C = \frac{P}{B},$$

where $B = \Sigma B_i$.

Thus, these cases refer strictly to production resulting from individual growth. The term q_1 expresses the specific rate of weight increase in mollusks during the first year of life; this index indicates to some extent the upper limit of specific production of the given species (in extreme situations where the population consists entirely of individuals aged less than one year).

The values of C for 23 species are indicated below (for half the species, C was calculated for the first time).

Mytilus galloprovincialis Lam. (Black Sea).
Calculations were based on data given in units of fresh weight.
For calculating q_1, the initial weight is taken as weight of ripe
eggs of 0.05–0.07 mm diameter (Vorob'ev, 1938); this amounted
to $1.2 \cdot 10^{-4}$ mg. Data on the weight increase of mussels on the
Odessa bank were used (Ivanov, 1967). Details on the age structure
of the population have been published by Ivanov (1965, 1968). The
main fraction (according to weight) consisted of 2-year-old
specimens (40–45%). Data on the biomass of the young forms are
lacking. It is known for M. edulis that specimens less than one
year old are scant in Danish waters (Smidt, 1951). I determined
that $C = 0.0025$.

Data on the weight increase of mussels during the first 3 years
of life can also be found in the work of Slavina (1965). Compared
with Ivanov's results, the data of Slavina yield higher values of q_i
during the first year of life but lower ones in the 1^+ and 2^+ year
groups. Since the first year group accounts for a negligible part
of the biomass, C must be lower than 0.0025 according to Slavina.
Dragoli (1966) indicated that this species usually lives 7–9 years,
rarely 12 years, and that its growth in many Black Sea biotopes is
much slower than assumed in my calculations. Thus, the value of
C which I calculated is probably closer to the upper limit for this
species.

Acmaea testudinalis (Müll.) (Barents Sea). Matveeva
(1955) studied the biology of this species in the Eastern Murman
area. She indicated the fresh weight of different age groups, as
well as the biomass and age structure of the population in four
biotopes. I used the data on the Yarnishnaya Gulf (where the
mollusks have the highest growth rate) and Dalnie Zelentsy inlet
(at Cape Povorotnyi where growth is slowest) and obtained the
following results: Yarnishnaya Gulf — $B = 36$ g/m^2, $C = 0.0011$,
$P_{ann} = 14.4$ g/m^2; Cape Povorotnyi — $B = 15.4$ g/m^2, $C = 0.0013$,
$P_{ann} = 7.2$ g/m^2. The specific production rate is slightly higher
at Cape Povorotnyi where mollusks grow more slowly. This is so
because mollusks living in the area indicated have a lower value
of q during the first year and higher values of q at the older ages
compared to the Yarnishnaya Gulf population. At the same time,
young forms were not considered in the calculation of C (since
their biomass was very small it was not indicated). The highest
recorded age of this mollusk is 6–7 years.

Acmaea digitalis Eschscholtz (Pacific Ocean, Oregon).
Frank (1965) examined the biology of this species. The life
span is 6–8 years; although the maximal length of 24–25 mm
is similar to that of A. testudinalis, this species grows more
rapidly. No weight data are available, but the volumes of individuals

of different sizes were determined. Assuming that the specific
weight of the mollusks changes only slightly with age (according
to Slavina, 1965, the specific weight of the Black Sea Mytilus
increases by only 13% while the weight changes from 15 to 70 g),
I calculated C in units of volume regarded as proportional to weight.
For the August age composition of the population a diurnal specific
production of 0.0035 is obtained.

Mytilaster lineatus (Gmel.) (Sea of Azov). Vorob'ev
(1949) examined the biology of this and the next two species. The
weight data are expressed in units of fresh weight. Vorob'ev
calculated the production of these 3 species for a period of 7 months
by comparing the increment and elimination of individuals and
determined the P/B coefficients by referring the 7-month production
to the minimal (spring) biomass. Greze (1967b) calculated the
value of C on the basis of Vorob'ev's data. I repeated these calculations by determining q_i from growth data. Vorob'ev indicated
the age structure of the population during the spring and fall.
Accordingly, two values of C were obtained, and their mean was
taken as the annual average. The values obtained for C for all three
species of mollusks are practically identical with those obtained
by Greze.

The maximal life span of M. lineatus in the Sea of Azov is
3 years. The following values were obtained: $q_1 = 0.025-0.034$,
$C = 0.005$ in spring, 0.014 in autumn; the annual average of
$C = 0.0095$. Greze found that $C = 0.0089$. According to Vorob'ev,
the production for a period of 7 months equals 901 g/m^2. If the
ratio between this value and average biomass (440 g/m^2) is divided
by the time (in days), the result will be $C = 0.010$. Thus, different
calculation methods yield similar values of C in this case.

Cardium edule L. (Sea of Azov). The life span of this
mollusk is 5 years, $q_1 = 0.033-0.040$. C equals 0.0018 in spring,
0.0064 in autumn. The annual average of $C = 0.0041$. Greze found
that $C = 0.0046$. According to Vorob'ev, the production for 7 months
equals 1,148 g/m^2 at an average biomass of 512 g/m^2. This leads to
$C = 0.010$. The value of C obtained from the production calculated
by Vorob'ev is twice that determined by other methods.

Abra (syndesmia) ovata (Phil.) (Sea of Azov). The
maximal life span of this mollusk is 4 years, $q_1 = 0.032-0.039$.
$C = 0.0018$ in spring and 0.0080 in autumn with 0.0049 as an annual
average. Greze found that $C = 0.0058$. According to Vorob'ev, the
production during 7 months equals 377 g/m^2 and the average biomass
is 235 g/m^2. Hence, we derived $C = 0.0080$.

Rissoa splendida Eichw. (Sea of Azov). Greze (1967b)
indicated that $C = 0.0106$ for this species without giving the source
of the initial data. According to Makkaveeva (1959b), this species

completes its life cycle in a year in the Black Sea; from her data I found that $q_1 = 0.032$.

Unionidae (Thames). Negus (1966) examined the biology and production of 3 unionid species. Biomass and production are expressed in units of dry weight of the soft tissues (without shells). According to these data, Unio pictorum L. has $C = 0.00038$, $q_1 = 0.02-0.03$, and a maximal life span of 13 years. For Unio tumidus Retz., $C = 0.00035$, $q_1 = 0.02-0.03$ and the maximal life span is 11 years.

Anodonta anatina L. $C = 0.00054$, $q_1 = 0.02-0.03$ and maximal life span is 10 years.

For calculating q_1 the weight of mature glochidia of A. anatina (0.066 mg) was taken as initial weight of all three species.

Dreissena polymorpha Pall. (Kuibyshev reservoir). Material necessary for calculating production is found in the work of Lyakhov and Mikheev (1964). Growth is expressed in units of fresh weight. I took the initial weight of postveligers ($2.5 \cdot 10^{-3}$ mg) as determined from their size (Kirpichenko, 1964) and found that $q_1 = 0.035$. Lyakhov and Mikheev have given the size structure of populations in many sections of the water body. I calculated C for 12 sectors and found similar values: $C = 0.0011-0.0017$ (average 0.0014). Mollusks less than a year old obviously constitute a negligible part of the population in terms of weight. This is corroborated from data on numbers of young (Kirpichenko, 1964). Because samples were collected in the seventh year of existence of the reservoir, the maximal age recorded is 6 years. The average biomass of this species on flooded forest soils was 1.8 kg/m^2; hence, for $C = 0.0014$; P_{ann} of 920 g/m^2 is obtained.

Sphaerium corneum L. (Belorussian lakes). The production of this species was examined by Arabina (1968), who found $C = 0.0044$. This form lives less than a year.

Bithynia tentaculata L. (Belorussian lakes). According to Arabina, $C = 0.0049$. The form lives up to 2 years.

Anisus vortex (L.), Gyraulus albus (Müll.), Valvata pulchella Stud. (Rybinsk reservoir). Tsikhon-Lukanina (1965 a, b) published growth curves and data on changes in the population biomass and average individual weight of these three species from May to September. These forms live less than a year. Calculations indicate that specimens weighing 2–4 mg have $q_w = 0.07-0.11$, and those weighing 8–10 mg have $q_w = 0.014-0.037$. Since the average individual weight in summer rarely exceeds 10 mg, the specific production is probably high: $C = 0.02-0.04$ for May–September or 0.01–0.02 as annual average (approximation).

Adacna vitrea (Ostr.) (Caspian Sea). Adding up the increment of existing biomass and biomass loss for the April–October period, Osadchikh and Yablonskaya (1968) arrived at a production of 137 kg/ha with an average biomass of 23.6 kg/ha. Assuming the mollusks do not grow during the rest of the year, $C = 0.016$. The maximal life span is a year.

Lacuna pallidula Da Costa (Barents Sea). The annual productions of this and the following species were calculated by Kuznetsov (1948 a, b), who compared the gain and loss of biomass in the Dalnie Zelentsy area. Kuznetsov determined P/B by relating production to initial biomass. For calculating C, I divided the production by the average biomass.

The annual production of L. pallidula in the Dalnie Zelentsy inlet equals 289 g/m^2. The annual average biomass is 25 g/m^2; hence $C = 0.03$. The life span of this species here is less than a year.

Margarita helicina (Phipp.) (Barents Sea). Kuznetsov (1948 b) calculated the production in several sectors of the Dalnie Zelentsy area. The life span is 20 months. $C = 0.008$ was obtained for the population in the Dalnie Zelentsy inlet and 0.020 for the Yarnyshnaya Gulf population.

Spisula elliptica Brown (Plymouth). Calculations are based on data expressed in units of total dry weight (Ford, 1925). This form does not live for more than a year, the age structure of the population changes steadily during the year, i. e., the popula- tion grows older between summer and winter. At the same time the value of C decreases. I calculated the following values of C: 0.080 on 5 July, 0.036 on 25 July, 0.016 on 19 September, 0.011 on 4 October, 0.05 in December—February. The annual average diurnal specific production (C) is 0.020.

Modiolus demissus Dillwyn (Atlantic coast of Georgia, U.S.A.). For the average biomass of 4,110 mg/m, the annual weight increment was estimated at 445 mg dry weight of soft tissues per m^2 (Kuenzler, 1961). However, it is noted that the increment can be underestimated by half. The approximate value of $C = 0.0003-$ 0.0006. The life span is not indicated, but the mollusks evidently live 7—8 years at least.

Psephidia lordi Baird (Pacific Ocean, Washington). Lie (1968) published a curve of linear growth based on peak dis- placement in age structure graphs. The relation between size and weight is also indicated. From this material, I prepared a weight increase curve and calculated the values of q_w. Of the age struc- ture graphs published, the most divergent ones (February and July 1963) were selected. The corresponding values of C are 0.003 and 0.009. The life span is 1.5 years.

Axinopsida sericata Carpenter (Pacific Ocean, Washing- ton). The initial data originate from the same publication as the preceding species. *C* equaled 0.006 in January 1963 and 0.004 in August 1964.

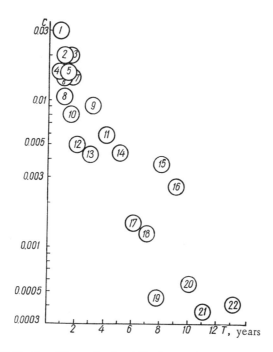

FIGURE 10. Specific production (*C*) of mollusks as function of their life span (*T*):

1) Lacuna pallidula; 2) Spisula elliptica; 3) Marga-rita helicina; 4)–6) Anisus vortex, Gyraulus albus, Valvata pulchella; 7) Adacna vitrea; 8) Rissoa splendida; 9) Mytilaster lineatus; 10) Margarita he-licina; 11) Abra ovata; 12) Bithynia tentaculata; 13) Sphaerium corneum; 14) Cardium edule; 15) Ac-maea digitalis; 16) Mytilus galloprovincialis; 17) Dreissena polymorpha; 18) Acmaea testudinalis; 19) Modiolus demissus; 20) Anodonta anatina; 21) Unio tumidus; 22) Unio pictorum.

A comparison of specific production data on different mollusks suggests that *C* is related inversely to the individual life span. This relationship can be demonstrated by an analysis of pertinent data (Figure 10). It was further assumed that mathematical models can be used to predict the exact nature of this relationship in mollusks (Zaika, 1970 b). Such models have indeed been constructed and can be found to have a certain relation to the empirically established dependence of the two indexes (see Chapter VI).

In the mollusks examined the value of C ranges from 0.0003 to 0.03. The production by reproduction (the total weight of eggs laid and the increment of the mollusks until settling on the ground) accounts for a relatively small part of the total production. According to Negus (1966), the production of glochidia in unionids represents only 10% of the production by growth without the shell. Clearly, the contribution of the glochidial production would be considerably smaller in terms of total fresh weight. Moreover, if the weight of the egg is taken as initial weight in calculating q_1, the entire production can be determined since the weight of progeny is reflected in the weight of the nongrowing mollusks.

6. SPECIFIC PRODUCTION OF ECHINODERMS

The productivity of echinoderms was hardly studied until recently. Not long ago the calculations of production of the starfish Asterias forbesi were published (Richards and Riley, 1967). Data necessary for calculating the specific productivity of three other echinoderm species have also been found.

Asterias forbesi (Desor) (Long Island). The specific production was calculated by Richards and Riley (1967). After metamorphosis the larva has a length up to 3 mm long. During the first year of life the starfish reach a length of 65 mm, and those over a year measure 66–125 mm. The weight increase ceases in winter; sometimes the weight even decreases due to worsening of food conditions. The specific production (C) is 0.023.

Asterias rubens L. (Plymouth). From data of Vevers (1949) I obtained a curve showing the linear growth of this species (length is defined as the distance from the mouth to the end of the largest arm). From these data the values of q_l and afterwards q_w (assuming that $q_w = 3 q_l$) were calculated. Vevers provides material on the average sizes of this species during different seasons. Since the average size usually ranges from 8.5 to 12 cm and calculating the respective values of q_w, I determined the specific production (C) at 0.01–0.02. This value is of course a rough estimate.

Amphoidia urtica (Lyman) (Pacific Ocean, Washington). Lie (1968) published a linear growth curve of this ophiurid based on peak displacement in size structure graphs. The relation between size and weight was also indicated. From these data I obtained a weight growth curve (in units of fresh weight), and by removing from it the values of specific growth increments q_w, C was determined for the age structure found in May and July 1963 (the months in which populations differed most in age structure). The respective values of C are 0.0022 and 0.0017. The maximal life span is about 5 years.

Cucumaria elongata (Northumberland). The biology of this holothurian was examined by Fish (1967). From the published data, I calculated q_w for individuals of different weights, and the values of C for the age structure indicated by the author. The specific production (C) is 0.0007. The maximal life span is not less than 10 years.

7. BIOLOGY AND SPECIFIC PRODUCTION OF THE APPENDICULARIAN OIKOPLEURA DIOICA IN THE BLACK SEA

Only one appendicularian species, Oikopleura dioica Fol., inhabits the Black Sea. I examined certain aspects of the biology of this species with special reference to seasonal changes in density and production indexes (Zaika, 1966, 1969 a). Analysis of data collected by the Institute of Southern Seas for many years on the abundance and distribution of appendicularians in the Black Sea shows that these animals reach a considerable density, especially in the neritic zone. After processing with the weighted moving average method, data on seasonal dynamics of appendicularian populations at different points along a 10-mile section in the Kamyshovaya inlet area indicate basic features of the seasonal changes in numbers. Figure 11 shows that these curves vary only slightly from year to year, and show three peaks — in February, June and August.

The main reproduction periods are May–June and August–September. The rapid growth and maturation rates in summer prevent the animals from reaching a large size; specimens having a body length of more than 0.7 mm are not encountered between May and September. In winter males reach a length of 0.83 mm, the females 1.0 mm. The largest females collected by us contain up to 100 eggs. In summer females begin reproducing at a body length not less than 0.65 mm, and in spring not less than 0.8 mm. According to available information, the females die immediately after reproducing.

The Black Sea appendicularians feed mainly on Pontosphaera huxley and small flagellates. Together with T. M. Kovaleva, I found in the stomach contents of appendicularians the algae Exuviaella,

Cyclotella, Peridinium, Glenodinium, Gymnodinium, Dictyocha, Thalassiosira and Distephanus. The maximal size of particle swallowed is about 40μ.

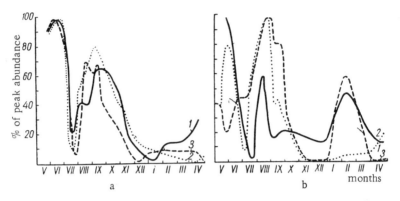

FIGURE 11. Seasonal changes of the relative numbers of Oikopleura dioica in the Sevastopol area:

1) 2.5 miles; 2) 7.5 miles; 3) 10 miles (initial data were treated by the weighted moving average method for 5 points); a) 1960–1961; b) 1961–1962.

TABLE 3. Relative numbers of different size groups of Black Sea appendicularians in samples collected during the summer of 1961

Group	Length, mm	Date						
		June		July			August		
		17	21	8	12	21	5	10	22
I	to 0.2	6	37	21	11	28	15	27	21
II	0.2–0.3	23	44	55	14	50	58	31	25
III	0.3–0.4	33	28	40	19	32	35	46	16
IV	0.4–0.5	30	27	28	14	22	23	33	18
V	0.5–0.6	2	12	10	7	8	19	11	5
VI	0.6–0.7	0	8	6	1	7	0	2	2
VII	0.7–0.8	0	1	0	0	3	0	0	0
VIII–IX	0.8–1.0	0	0	0	0	0	0	0	0
Total		94	157	160	67	150	150	150	87

Appendicularians often reach great densities and may consume considerable amounts of phytoplankton (Lohmann, 1899) and food suitable for young fishes (Shelbourne, 1962). Thus, studies of the productivity of these forms are very interesting. Because experimental data on the growth of appendicularians are lacking, I plotted a rough growth curve based on the size composition of the population.

The size composition of the population was determined from samples taken in 1961 (Table 3). Comparing these data with the dynamics of the total abundance of the appendicularians during the same period curves were obtained showing changes in the abundance of each size group. The groupings used were larger than those in Table 3 (the 9 groups of this table were combined as follows: I–II, III–IV, V-IX). These three major size categories have the following average body lengths: 0.2, 0.4 and 0.6 mm. The population dynamics curves of these size categories are shown in Figure 12.

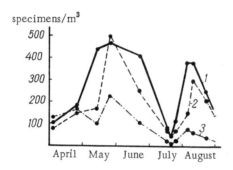

FIGURE 12. Seasonal changes of numbers of Oikopleura dioica in the neritic zone of the Black Sea in 1961, according to size groups:

1) I-II; 2) III-IV; 3) V-IX.

Using Figure 12, let us attempt to determine the time required for specimens measuring 0.2 to reach successively lengths of 0.4 and 0.6 mm. (Such estimates can only be regarded as initial approximations.) Abundance of group V—IX increased slightly in April and then decreased until mid-May, while while that of group I–II continued to increase. The decreased abundance of group V–IX during this period can be attributed to intensive reproduction which is accompanied by death of females (Fenaux, 1963). In May, the numbers of group V–IX began to increase. According to the curve of group I–II, the generation reaching a body length of 0.6 mm and a density of 230 specimens/m^3 on 24 May was 0.2 mm long on 1 May. Consequently, growth of specimens from 0.2 to 0.6 mm took not longer than 15–17 days, and from 0.4 to 0.6 mm about 8–9 days.

Because the total abundance decreases sharply each year in June—July, it is not possible then to determine the growth rate from changes in numbers of the different age groups. In August, the density peak of group V–IX is found in such a relation to the abundance

curves of groups I–II and III–IV that it supports the conclusion that
growth from 0.4 to 0.6 mm lasted 5–7 days and 0.2 to 0.6 mm from
10–12 days.

From these considerations, it was tentatively assumed that the
body growth from 0.2 to 0.6 mm lasts 15–17 days in May and
10–12 days in August. The relation between these values appears
acceptable considering that water temperatures are 14.8°C in May
and 24.6°C in August.

From these data a growth curve was constructed according to
the method described in Chapter V for the amphipod A c a n t h o -
g a m m a r u s .

According to available information (Bary, 1960) and my own
measurements, the maximal body length of O. d i o i c a is approximately 1 mm. Therefore I took l_∞ = 1 mm. Using the above-
mentioned analysis of growth of appendicularians in August, it was
found from equation (61) that α = 0.064. This allows a complete
description of the linear growth.

Petipa (1956) determined by weighing that appendicularians
having a body length of 0.586 mm weigh 0.014 mg. According to
the weight increase equation

$$w_t = w_\infty (1 - e^{-\alpha t})^3, \qquad (64)$$

knowing w_t = 0.014, α = 0.064 and t, w_∞ can be found. For this,
it is necessary to determine the value of t corresponding to a
weight of 0.014 mg, i. e., a length of 0.586 mm. This can be done
using equation (60); t = 13.8 days. Then w_∞ = 0.07 mg. The weight
increase curve can be constructed using equation (64).

TABLE 4. Production indexes of the Black Sea appendicularian O. d i o i c a during the summer
of 1960

Period		P, mg/m³	B, mg/m³	Average individual weight, mg	Diurnal c
June	first half	6.42	0.99	0.0011	0.43
	second half	6.30	1.08	0.0014	0.39
July	first half	4.74	0.94	0.0018	0.33
	second half	4.20	0.91	0.0020	0.30
August	first half	4.74	1.02	0.0019	0.31
	second half	5.28	0.94	0.0015	0.37
September	first half	5.88	1.16	0.0018	0.34
	second half	5.60	1.20	0.0020	0.31
Total	....	43.16	–	–	–
Average	....	–	1.03	–	0.35

The production of appendicularians was calculated using the method of Greze (second variant of the graphic method). Gonads are present in the hinder part of the body in appendicularians, and their size determines total body length. This condition, together with the fact that females die upon reproduction, makes determination of production by reproduction unnecessary because it is already taken into account. The weight increments were determined using the equation

$$\frac{dw}{dt} = 3\alpha \, (w_\infty^{1/3} w^{2/3} - w). \tag{65}$$

Calculations were made for 15-day intervals for the summer of 1960 and for 10-day intervals for the summer of 1961. Data on production of appendicularians in the summer fo 1960 are presented in Table 4.

The main indexes of the production process of appendicularians during the next two summers were as follows:

	1960	1961
Average biomass, mg/m^3	1.08	1.21
Production per season, mg/m^3	43.16	41.0
Diurnal specific production (average for the summer)	0.34	0.32

It is necessary to note that although the calculation of appendicularian production was based on a rough estimate of the growth rate, the size structure of the population is known accurately. Therefore I extensively used results of determinations of C in appendicularians for analyzing the dependence of specific production on the age structure of the population. The calculations indicate that if there is an error in the value of α then the absolute values of C and P would be affected but not the ratio between values of C obtained for different age structures. Thus, the main considerations concerning appendicularians presented in Chapter VI remain valid regardless of the value of α. The same is true of sagittas.

8. GROWTH AND SPECIFIC PRODUCTION OF SAGITTA SETOSA IN THE BLACK SEA

Together with studies of productivity of appendicularians, a similar study was made of Sagitta setosa from material collected in the same area (Kamyshovaya section). Studying the size composition of sagittas in the samples, the peak displacement in graphs of the size structure of the population was analyzed.

In this way, pairs of l_1 and l_2 values were obtained together with the respective Δt which were used for determining α after equation (61). The following series of values was obtained:

l_1, mm	l_2, mm	Δt, days	α
3.5	5.5	4	0.026
2.5	3.5	2	0.025
4.5	5.5	2	0.025
3.5	5.5	2	0.055
1.5	2.5	2	0.024

All these cases pertain to the summer (average water tempera-ture of 22°C). The penultimate value was discarded because it sharply diverges from the general series, and $\alpha = 0.025$ was taken as the summer average. From existing data on Black Sea sagittas it was assumed that $l_\infty = 23$ mm. Since the concentration of sagittas per m^3 is very low, calculations of diurnal and specific production were made per 100 specimens with reference to the actual age structure of the population during each 15-day period. The production per m^3 was then determined. The following indexes were obtained:

	Summer, 1960	Summer, 1961
Production per season, mg/m^3 ..	269	264
Diurnal specific production (summer average)	0.21	0.31

Production calculations were based on individual growth only. Although the sagittas are large predators, they nevertheless had a high production. Since my calculations were based on a rough estimate of the growth rate, the data obtained had to be verified by experimental study of sagitta growth.

Despite considerable difficulties, Mironov (1970) obtained 11 positive results in experiments on the composition of sagittas. The values of l_1, l_2 amd Δt obtained yield a complete growth curve of sagittas (see above):

Δt, days	5	15	27	6	8	2	9	2	6	7	2
l_1, mm . . .	3.1	5.5	7.0	5.2	4.0	4.0	4.0	4.0	4.0	5.2	5.8
l_2, mm . . .	4.0	7.7	12.5	6.7	7.0	4.8	6.1	4.4	5.7	6.6	6.8

The average temperature during the experimental period was 14.7°C. Calculation of α using equation (61) for these data gave values rather evenly dispersed in the interval of α = 0.009—0.021, except for α = 0.030 obtained in the last experiment. Excluding the results of the last experiment, an average of α = 0.014 is obtained (including last experiment, α averages 0.015).

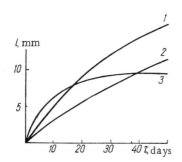

FIGURE 13. Linear growth curves of sagittas:

1) Sagitta setosa (at 22°C), after original data; 2) S.Setosa (at 14.7°C), after Mironov (1970); 3) S.hispida, after Reeve (1968).

The linear growth curve of S. setosa according to data of Mironov (1970) differs from that obtained from my own data (Figure 13), since one refers to a temperature of 14.7°C and the other to 22°C. The only procedure (and not a reliable one at that) at our disposal for comparing the two curves involves a conversion based on the "normal curve" of Krogh (Vinberg, 1956). Such a conversion is used in growth studies on the assumption that changes in growth rate are related to temperature like those in respiration oxygen uptake. According to equation (65), characterizing the rate of weight increase, the growth curve at 22°C can be converted to one at 15°C by multiplying α = 0.025 by the corresponding coefficient because the growth rate at any age changes by a magnitude that can be determined by a conversion rule based on the Krogh curve. Thus for 15°C we obtain α = 0.013. As shown earlier, the data of Mironov give a very similar value (α = 0.014). Consequently, Mironov's experimental data (1970) are practically identical with my results which were obtained indirectly by calculating the growth curve with correction for temperature. This is confirmation of my production calculations for sagittas.

Finally, it would be interesting to compare the material presented with that obtained by Reeve (1968) in cultures of Sagitta hispida. Several specimens of this species were successfully grown from 5.5 mm to 9 mm at 21°C. According to my calculations, the growth results agree well with the Bertalanffy equation for $l_\infty = 9.7$ mm and $\alpha = 0.010$. Thus, the definitive size of S. hispida is much smaller than in S. setosa but its growth rate is much higher (see Figure 13).

Estimates are now available of specific productions of more than 100 species of freshwater and marine invertebrates of different groups. Although more accurate estimates are required to solve certain theoretical and practical problems in a large number of species, the existing data provide a rather solid foundation for seeking quantitative relationships between specific production and other indexes. This material also provides a general idea of the level of the productivity of various groups of aquatic animals.

Chapter VI

PRINCIPLES DETERMINING THE SPECIFIC
PRODUCTION OF THE POPULATION

Chapters IV and V present the values of specific productions of many species of aquatic animals, starting from infusorians (and, partly, bacteria) up to higher forms. However, the respective values for many species have to be determined more accurately and the fluctuations studied (seasonal and multiannual changes of C, etc.); moreover, some major taxonomic and ecological groups are still uninvestigated.

Research on productivity, especially specific production, cannot be restricted to the recording of values of C for an increasing number of species. It is exceedingly important to seek and analyze the principles which determine the productivity level of different species, that is, to develop a theory of productivity. Qualitative aspects of the relation between productivity and certain biological features of the population or the individuals have been long known. It was known, for instance, that the production of a population increases with the growth rate of individuals of the given species; the same applies to the effect of temperature on the productivity of poikilothermic animals. I believe the basic problems in the analysis of the factual data to be the search for and establishment of quantitative relations between productivity and various populational and environmental factors. Material accumulated so far is quite sufficient for the solution of such problems.

The present chapter deals with results of my own investigations in this direction. The standard approach was to compare the factual data and to seek quantitative relationships between specific production and parameters interesting to us. If a quantitative principle was discovered, the same empirical approach served as the basis for seeking an explanation of the observed phenomenon. Mathematical models played a major role in this work. Since specific production is a calculable quantity, the value of C can be calculated without any difficulty for any theoretical case from known biological principles relating to the derivation of calculated values and by using logical assumptions. Such a model indicates the quantitative relations of the indexes investigated. Comparison with factual data improves the model, which will consequently yield more correct results. The use of models sheds light on the existing relationships and explains the "inner mechanism" of their formation, as well as indicating some crucial problems and sometimes ways for their solution.

1. SPECIFIC PRODUCTION AND AGE STRUCTURE OF THE POPULATION

As a rule, the specific rate of weight increase declines with age. This leads to lower specific production in populations containing a greater proportion of older age groups. The age structure of many populations continuously changes and this markedly influences the specific production. Hence quantitative assessment of the effect of age structure on specific production in the population is definitely interesting. This can be facilitated by finding a sufficiently simple index for characterizing the age structure of the population.

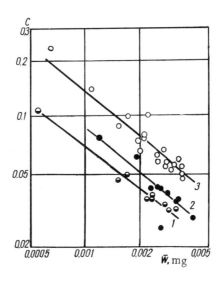

FIGURE 14. Relation of specific production (C) to average individual weight ($\bar{w}$) of A c a r t i a   c l a u s i at different temperatures:

1) 16; 2) 19; 3) 23 (logarithmic scales).

Assuming that the average individual weight ($\bar{w}$) may serve in many cases as an index of the age structure of the population, this quantity was compared with the diurnal specific production over a period of 10–15 days in some representatives of the zooplankton of the Black Sea and Sea of Azov (Zaika and Malovitskaya, 1967). The data were taken for periods of equal average water temperature. Figures 14–17 show results in lines drawn through points on logarithmic scales. The dependence of C to $\bar{w}$ has the form

$$\log C = a - b \log \bar{w},$$

where a and b are coefficients.

For A c a r t i a   c l a u s i (Sea of Azov) points are plotted for three different temperatures. The position of the line obviously depends on temperature. This is because the individual growth rate increases with temperature.

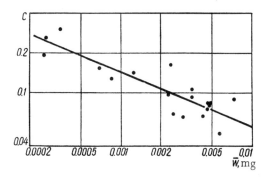

FIGURE 15. Relation of specific production (C) to average individual weight $(\bar{w})$ of C e n t r o p a g e s   k r o y e r i (logarithmic scale)

Thus, a rather simple quantitative relationship exists between C and $\bar{w}$. Since the average individual weight is calculated simply (by dividing biomass by the number of individuals), it would be desirable in further studies to examine the use of $\bar{w}$ as an index of the age structure and as a parameter linked in a definite manner to specific production.

The following model was proposed to explain the observed quantitative relationship between C and $\bar{w}$ (Zaika, 1968; Zaika and Andryushchenko, 1969). The infinite variety of concrete cases of age structure had to be classified into a few general, idealized types. The populations of many species usually consist of individuals of practically equal age. Such a condition is observed in populations having a brief period of mass reproduction and whose individual life span (maximal) is roughly equal to the time between two

reproduction peaks (the parental generation dies soon after reproduction). Such populations have the "simplest structure" according
to the terminology of Bekman and Menshutkin (1964) who published a
mathematical analysis of their productivity.

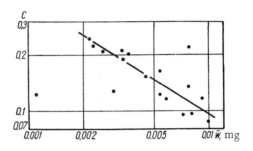

FIGURE 16. Relation of specific production (C) to average individual
weight ($\bar{w}$) in C a l a n i p e d a   a q u a e - d u l c i s (logarithmic scale)

In other, more widespread cases, the population contains many or all
the age groups. Such populations show considerable variations in the
proportions of the different age groups. Two basic situations can be
distinguished: 1) young ages predominate; 2) old ages predominate.

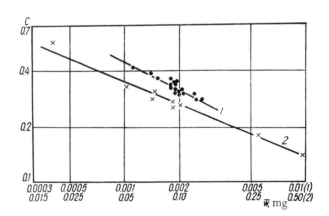

FIGURE 17. Relation of specific production (C) to average individual
weight ($\bar{w}$) of O i k o p l e u r a   d i o i c a (1) and S a g i t t a   s e t o s a (2)
(logarithmic scales)

Let us first examine the specific production of populations of the simplest structure. Since at any given moment these populations consist
of individuals of the same age, their specific production corresponds to

specific growth rate of individuals of the given weight. If limited to species whose growth is adequately covered by the Bertalanffy growth theory, the individual growth rate can be expressed by the equation

$$\frac{dw}{dt} = K \, (w_\infty^{1-\alpha} w^\alpha - w), \tag{66}$$

where w_∞ is the theoretical limit of growth in the given conditions, and α and K are coefficients.

The specific rate of weight increase is

$$q_w = \frac{dw}{dt} \cdot \frac{1}{w} \, ,$$

or, taking into account equation (66),

$$q_w = K \left(\frac{w_\infty^{1-\alpha}}{w^{1-\alpha}} - 1 \right). \tag{67}$$

As already noted, the specific production of the population of simplest structure is related to variations in individual weight. Using equation (67), a curve was drawn relating C to w (Figure 18 a) using the frequently used value of $\alpha = \frac{2}{3}$ (Vinberg, 1966).

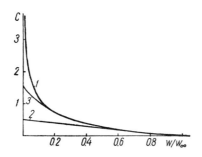

FIGURE 18. Specific production of the population (C) as a function of average individual weight (expressed in fractions of w_∞):

1) population of simplest structure; 2) population containing individuals not weighing more than w/w_{max}; 3) individuals of all ages are present.

In further discussion, the specific production of populations of the simplest type will be designated as C_0. Obviously, C_0 varies inversely with individual age (weight) and reaches the maximal possible value for minimal $\bar{w}$. Since the position of the curve depends on the temperature, the effect of age structure on the specific production of the population will be treated here and below as though temperature is constant.

While in populations of the simplest structure the specific rate of individual growth and the specific production (which is equal to the former) depend (at constant temperature) solely on the individual weight, in more complex populations C depends in addition on the distribution $N_{(w)}$ of abundance by weight ($N_{(w)}$ is the abundance of individuals with weight ranging from w_0 to w).

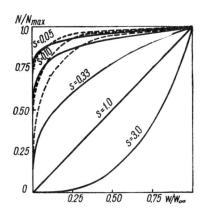

FIGURE 19. Abundance distribution by weight (cumulative abundance) for Oikopleura dioica.

Continuous lines represent theoretical, dotted lines actual functions.

The age structure of populations of zooplankton in the material used for this work depends on a number of interrelated factors which will not be discussed here. To simplify the model (and, to some extent, at the expense of biological interpretation) it shall be assumed that the common pattern of weight distribution in many zooplanktons can be roughly represented by a simple function of the type

$$N_{(w)} = N\left(\frac{w}{w_{max}}\right)^s ,\qquad (68)$$

where N is the total size of the population and w_{max} the maximal individual weight in the population at a given moment (unlike w_∞, which is the maximal theoretically possible individual weight for the same population).

Equation (68) can be written in the form

$$\frac{N_{(w)}}{N} = \left(\frac{w}{w_{\max}}\right)^s.$$ (68')

Obviously $\frac{N_{(w)}}{N}$ indicates the proportion of individuals whose weight ranges from w_0 to w (or from 0 to w if initial weight is negligible compared to that of adults). Both sides of equation (68') contain dimensionless quantities, which makes it possible to compare the functions of weight distribution in different species or different states of the same population, regardless of the absolute values of numbers and weights.

Giving s values from 0 to ∞, a series of curves is obtained, some of which are shown in Figure 19.

When the population has an even weight distribution, $s=1$; if $s<1$, individuals of low weight predominate while if $s>1$ the individuals are of great weight. Naturally, the actual distributions of abundance by weight may differ somewhat from the model, especially if the population lacks certain intermediate age groups. In usual practice, however, researchers determine the abundance distribution by weight in terms of major age categories (for example, naplii, copepodids and adults). The resulting curves are smooth and often nearly parabolic. For example, Figure 19 shows curves of abundance distribution by weight for the Black Sea population of O i k o p l e u r a d i o i c a at different times in the summer (dotted lines). Calculation of C from these actual curves gave values slightly exceeding those obtained using close parabolas.

These assumptions on individual growth and population structure allow determining specific production with reference to an actual population structure reflected in the value of s. The production of the population is defined by the expression

$$P = \int_0^{w_{\max}} \frac{dw}{dt} n\,dw,$$ (69)

where n is the density of the distribution of abundance by weight, which, by equation (68), gives

$$n = \frac{dN}{dw} = \frac{sN}{w_{\max}} \left(\frac{w}{w_{\max}}\right)^{s-1}.$$ (70)

Substituting $\frac{dw}{dt}$ and n into equation (69) from equations (66) and (70), we obtain after integration

$$P = KN\left(\frac{s}{s+\alpha}\, w_\infty^{1-\alpha} w_{\max}^\alpha - \frac{s}{s+1}\, w_{\max}\right).$$ (69')

The population biomass is expressed by the equation

$$B = \int_0^{w_{\max}} wn\,dw = \frac{s}{s+1} N w_{\max}.$$ (71)

Then

$$C = \frac{P}{B} = K\left[\frac{s+1}{s+\alpha}\left(\frac{w_\infty}{w_{max}}\right)^{1-\alpha} - 1\right]. \tag{72}$$

In this case the average individual weight $\bar{w}$ in the population can also be expressed in terms of s:

$$\bar{w} = \frac{B}{N} = \frac{s}{s+1}w_{max}. \tag{73}$$

It is obvious that $0 < \bar{w} < w_{max}$ at $0 < s < \infty$. Equation (72) links C and s. By means of equation (73), C can be rendered as a function of $\bar{w}$, which is an easily available index:

$$C = K\left[\frac{w_{max}}{\alpha w_{max} + (1-\alpha)\bar{w}}\left(\frac{w_\infty}{w_{max}}\right)^{1-\alpha} - 1\right]. \tag{74}$$

Finally, for comparison with the behavior of C_0 (see Figure 18) $\bar{w}$ can be expressed in equation (74) as a fraction of w_∞:

$$C = K\left[\frac{1}{\dfrac{\alpha w_{max}}{w_\infty} + (1-\alpha)\dfrac{\bar{w}}{w_\infty}}\left(\frac{w_\infty}{w_{max}}\right)^{-\alpha} - 1\right]. \tag{75}$$

Equation (75) makes it possible to plot curves of C according to $\dfrac{\bar{w}}{w_\infty}$ (see Figure 18, 2 and 3). Here it is assumed that $\alpha = \dfrac{2}{3}$, as in the case of an individual (Figure 18, 1). Curve 2 in Figure 18 refers to cases in which $\dfrac{w_\infty}{w_{max}} = 1$, i. e., in populations consisting of individuals of all weights ranging from 0 to w_∞. Obviously, the curve of C is situated below the curve of C_0, i.e., C of a population having an average individual weight w is always lower than that of an individual which weighs w; with decreasing w the specific production C rises more slowly than C_0.

Curve 3 in Figure 18 refers to cases in which $\dfrac{w_\infty}{w_{max}} > 1$. In such cases if $\bar{w} = w_{max}$, the specific production C equals C_0 for $w = w_{max}$. If the population has age structures characterized by different values of both s and $\dfrac{w_\infty}{w_{max}}$ in different periods, the values of C must be distributed in the area between the curves of C (for $\dfrac{w_\infty}{w_{max}} = 1$) and C_0.

Naturally, these limitations apply only in conditions preserving a constant temperature and employing equation (68).

Actually, only a few data can be used for comparison with the model. For this purpose, values of C are needed from a single population at different age structures at the same temperature. Suitable data for several species were selected and compared with the theoretical curves (Figure 20). Since the growth curves of two species (Oikopleura dioica and Sagitta setosa) were constructed assuming that $\alpha = \frac{2}{3}$, all the theoretical curves in the figure were checked for the same condition.

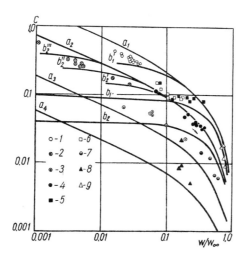

FIGURE 20. Specific production (C) as a function of age structure of the population (expressed as the ratio w/w_∞):

The lines $a_1 - a_4$ refer to populations of the simplest structure; the lines b_1 and b_2 to the case $\frac{w_\infty}{w_{max}} = 1$; the lines b_1', $b_2' - b_2'''$ to the cases when $\frac{w_\infty}{w_{max}} > 1$;

1) Oikopleura dioica; 2) Sagitta setosa; 3) Orchestia bottae; 4) Acartia clausi (Sea of Azov); 5) A.clausi (Black Sea); 6) Centropages kroyeri; 7) Arctiodiaptomus bacillifer; 8) Acanthodiaptomus denticornis; 9) Daphnia longispina (Zaika, 1968).

Four curves of C_0 ($a_1 - a_4$) were obtained; a_1 refers to a case of $K = 1.92$ and a_2 to $K = 0.78$. These are values of K characterizing the growth of O. dioica and S. setosa, respectively (a_3 and a_4 are likewise C_0 curves characterizing the specific production of populations of the simplest structure, but at lower values of K).

The curves $a_1 - a_3$ correspond to the curves $b_1 - b_3$ describing changes in C in the population when $\frac{w_\infty}{w_{max}} = 1$. Since O. dioica and S. setosa actually had, during the experimental period, certain definite values of the indexes $\frac{w_\infty}{w_{max}} > 1$, additional curves ($b_1'$, $b_2' - b_2'''$) were prepared for the corresponding values of $\frac{w_\infty}{w_{max}}$.

It can be seen in the example of these species that the values of C calculated from actual data and plotted on $\bar{w}$ lie above the corresponding theoretical curves of C but below the C_0 curves. This is due, as proven by the control, to the deviation of the actual curves of weight dis-tribution from the parabolic. It may be assumed that if $\frac{w_\infty}{w_{max}} = 1$ the curves of C indicate the lower limit of C, especially at small values of $\bar{w}$. In the case of S. setosa it is seen that decrease in the values of $\bar{w}$ is accompanied by a rise of the $\frac{w_\infty}{w_{max}}$ ratio.

Thus the relation between C and $\bar{w}$ can be expressed by the equation $\log C = a - b \log \bar{w}$ within a given range of $\bar{w}$ values; in general cases the form is different as described by the proposed model for the relation between C and $\bar{w}$. For other species described in Figure 20, only the general distribution of the points can be appraised. Although Bertalanffy's growth theory was not used in constructing the growth curves, the distribution of weight in the population was not compared with the theoretical model and the production was calculated by different methods. Despite these reasons for deviation from theoretical expectations, the relative position of the points is rather close to that expected for the given $\bar{w}$. Determining the value of $\frac{w_\infty}{w_{max}}$ in each case, individual growth and weight distribution of these species might be described by selecting suitable values of K and s based on the position of C points in Figure 20.

The following general comments can be made after analysis of the model I have proposed to relate specific production of the population to its age composition as reflected by average individual weight. The upper limit of variations of C is the curve of C_0, which is inac-cessible with a complex age structure. At the same time the value of C_0 changes by factor of less than 10 with variation of $\frac{\bar{w}}{w_\infty}$ within 0.1 to 0.001. Irrespective of the form of the function of the weight distri-bution of the population, it can be assumed that the fluctuation of C as a function of $\bar{w}$ is smaller in populations containing a larger number of age groups.

In the special case defined by equation (68) and in the presence of all age groups $\left(\frac{w_\infty}{w_{max}} = 1\right)$, the specific production of the population remains unchanged for $0.001 \leqslant \frac{\bar{w}}{w_\infty} \leqslant 0.1$.

With an increased $\frac{w_\infty}{w_{max}}$ ratio and, generally, for simpler age com-positions approaching the simplest populational structure, the

variability of C increases and approaches C_0. The material
examined for several zooplanktonic populations shows however,
that hundredfold fluctuations of the value of $\bar{w}$ are not often observed.
The variability of C as related to the age structure is correspond-
ingly less.

    Thus, the proposed model provides a sufficiently complete
explanation of the relation between C and $\bar{w}$. If $\bar{w}$ varies within
relatively narrow limits, the relation between C and w can be
approximated by the simple equation log $C = a - b$ log $\bar{w}$.

2. SPECIFIC PRODUCTION OF THE POPULATION
AND WATER TEMPERATURE

    The effect of ambient temperature on the metabolic rates in
poikilothermic animals is well known. With respect to specific
production of a population, the analysis of the temperature effect
is especially difficult because several processes, each depending
differently on the temperature, must be considered. Actually, a
comparison of the state of the population during different months,
for example, shows fluctuations in individual growth and reproduction rates as well as in the age structure. All these variations affect
the magnitude of the diurnal specific production of the population,
indicating a link to temperature via the above-mentioned processes.

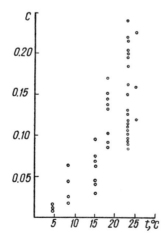

FIGURE 21. Relation of specific production (c) of C a l a n i p e d a
a q u a e - d u l c i s to water temperature (t, °C)

An empirical attempt was made to find a quantitative relationship between water temperature and specific productions for
populations of two planktonic copepod species in the Sea of Azov
(Zaika and Malovitskaya, 1967). The growth curves required for
determining the production of these species were made for only
a few usual temperatures instead of the whole range. As a result,
the values obtained are densely grouped with respect to temperature. Figure 21 shows all the values of C we obtained for C a l a n i -
p e d a   a q u a e - d u l c i s. The tendency for C to increase with
temperature is obvious. An arithmetic mean of C was found for
each temperature used in the calculation. Figure 22 gives these
mean values for two species (A c a r t i a   c l a u s i   and   C a l a n i -
p e d a   a q u a e - d u l c i s). It can be seen that in these cases the
relationship between C and water temperature has the form $\log C =$
$= p + q \log t°$, where p and q are coefficients.

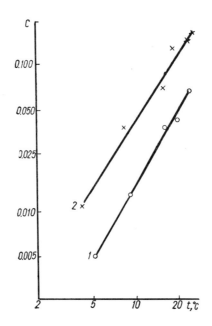

FIGURE 22. Relation between average values of specific production (c)
and water temperature (t, °C):

1) A c a r t i a; 2) C a l a n i p e d a (logarithmic scale).

As already noted, empirically established relationships between
C and $t°$ cannot be explained simply since they depend on a number
of functions. In particular, according to material presented at the
beginning of the chapter, specific production depends in a certain way on
the age structure of the population. In simple form, this relationship

can be conveyed by the expression log $C = a - b$ log $\bar{w}$ in which the average individual weight $\bar{w}$ serves as an index of the age composition. It is known that the average individual weight of the population decreases during intensive reproduction. If the biology of a given animal is such that the reproduction period coincides with the seasonal changes in water temperature, then, most probably, there is a definite relationship between temperature and average individual weight.

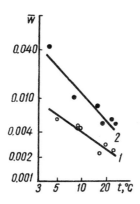

FIGURE 23. Relation between average individual weight ($\bar{w}$) and the water temperature (t, °C)

1) A c a r t i a c l a u s i; 2) C a l a n i p e d a a q u a e - d u l c i s.

The arithmetic means of $\bar{w}$ for A . c l a u s i and C . a q u a e - d u l - c i s at different temperatures are plotted in Figure 23. The function log $\bar{w} = m - n$ log t° is obtained, where m and n are coefficients. Although this formula is by no means universal, it reveals a concrete relation between $\bar{w}$ and t° and can be used for analyzing the relation between C and t° obtained for this particular animal species (see Figure 22).

The value of C also depends on the individual growth rate which in turn is influenced by water temperature. To calculate C for the planktonic copepods discussed in this section, growth curves for different temperatures were obtained as follows. In the case of C . a q u a e - d u l c i s it was assumed that the rate of development depends on temperature as established in the experiments of Kudelina (1950), who found that the development takes 60 days at 10°C, 30 days at 17°C, 18 days at 22°C, and 12 days at 26°C. Since the

growth rate $\frac{dw}{dt}$ is inversely proportional to the duration of

development the relationship between $\frac{dw}{dt}$ and $t°$ obtained from the
above data is $\log \frac{dw}{dt} = d + rt°$, where d and r are coefficients.

Growth curves of A. c l a u s i were obtained using corrections
based on the Krogh curve (Vinberg, 1956); results of this calculation agreed satisfactorily with the development periods of
different generations in the Sea of Azov (Malovitskaya, 1967). The
dependence of growth rate on temperature is also adequately

expressed by the equation $\log \frac{dw}{dt} = d + rt°$.

Due to limitation of the material, I did not attempt to construct
a single model which would take into consideration the special
relations between temperature and each of the processes involved
and also reveal theoretically the possible qualitative relation
between C and $t°$. In my view, the relation between $\overline{w}$ and $t°$ indicated above cannot have a general application. Clearly, the construction of the $C - t°$ model requires extensive information on the
nature of the dependence of $\overline{w}$ and $t°$. In particular, it is desirable
to obtain data on the variability of C as a function of $t°$ for different
species. This will help determine the applicability of the equation
$\log C = p + q \log t°$ which I established for two copepod species.
It is interesting that Sushchenya (1967) found a similar relationship
between C and $t°$ in the amphipod O r c h e s t i a   b o t t a e (Figure 24).

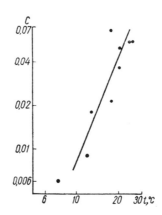

FIGURE 24. Relation between specific production (C)
of Or c h e st i a   b ott a e and temperature (t, °C)

The above presentation indicates that for 3 crustacean species
we now possess an empirically established relationship between
specific production and ambient temperature. Since the relation of
growth and development rates in animals to temperatures has been
studied in detail (Vinberg, 1968 a), particular attention should be paid

to collecting material which might explain the changes in age structure of the population in relation to temperature.

3. GROWTH RATE, LIFE SPAN AND SPECIFIC
PRODUCTION OF MOLLUSKS

Analysis of the productivity of many species of aquatic mollusks
indicated that the annual average diurnal specific production (C)
decreases the longer the given population lives (Zaika, 1970 b).
Relevant factual material is presented in Chapter V. The present
section is devoted to constructing and analyzing a mathematical
model demonstrating how the relations between individual growth
rate, maximal life span and age composition of the population
determine the value of specific production (Zaika and Ostrovskaya,
1971).

MATHEMATICAL EXPRESSION OF VALUES AND
PROCESSES USED IN THE MODEL

Individual growth rate. Many more published data are available
on linear growth in mollusks than on weight increase. Accordingly,
the following discussion will be based on linear growth. It is not
necessary here to analyze the different possible ways of describing
growth. It is sufficient to select a simple equation which satisfactorily expresses growth in mollusks. This is the Bertalanffy
equation of attenuating linear growth

$$l_\tau = l_\infty (1 - e^{-\alpha \tau}), \tag{76}$$

where l_τ is the size (length, height, etc.) reached by the individual
at age τ_1, and l_∞ is the theoretical limit of individual size (at $\tau \to \infty$).
The values of the parameters of equation (76) were approximated
using material on the growth of many species of aquatic mollusks.
This gave the opportunity to demonstrate that equation (76) can be
used to describe growth in most cases. In addition a rough estimate was made of the limits of the different parameters. The applicability of equation (76) is illustrated in Figure 25 describing growth
in some species of Mytilus.
According to equation (76), the growth curve is determined by
two parameters, α and l_∞ (Figure 26).

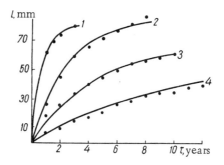

FIGURE 25. Linear growth of some mollusks:

1)−3) M y t i l u s g a l l o p r o v i n c i a l i s (1 − after Slavina, 1965;
2− after Ivanov, 1967; 3 − after Dragoli,1966); 4) M y t i l u s e d u l i s
(after Matveeva, 1948). The points are based on actual data; the
lines are theoretical from equation (76).

The specific rate of linear growth is

$$q_l = \frac{dl}{d\tau} \cdot \frac{1}{l} . \tag{77}$$

Comparison of equations (76) and (77) shows that q_w is a function
only of the parameter a:

$$q_l = \frac{\alpha e^{-\alpha\tau}}{1 - e^{-\alpha\tau}}. \tag{78}$$

The relation between linear growth and weight increase is evident
from the equation

$$w_\tau = bl_\tau^m, \tag{79}$$

where b and m are constants; since the value of m in the animals examined is usually about 3, $m = 3$ was taken in the model. Accordingly, the
following expression for specific rate of weight increase (q_w) is derived
from equations (76) and (79),

$$q_w = \frac{dw}{d\tau} \cdot \frac{1}{w} = \frac{3\alpha e^{-\alpha\tau}}{1 - e^{-\alpha\tau}} . \tag{80}$$

It is obvious that q_w, like q_l, depends on the value of the parameter
a. This is important because the equations used for calculating
specific production of the population include only the specific growth
rate which, as we see, does not depend on the absolute dimensions
of the animal. This rather unexpected conclusion requires explanation. Here I shall only note that this conclusion can be verified by
a detailed description of equation (88) (see below). The mathematical procedure involves cancellation of l_∞ in both the numerator
and denominator.

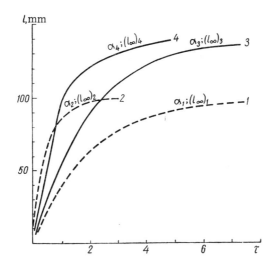

FIGURE 26. Individual linear growth curves (l — length, τ — age in arbitrary units at different values of α and l_∞):

1), 2) $(l_\infty)_1 = (l_\infty)_2$; $\alpha_1 < \alpha_2$; 3), 4) $(l_\infty)_3 = (l_\infty)_4$; $\alpha_3 < \alpha_4$; 1), 3) $\alpha_1 = \alpha_3$; $(l_\infty)_1 < (l_\infty)_3$; 2), 4) $\alpha_2 > \alpha_4$; $(l_\infty)_2 < (l_\infty)_4$.

Life span.   Life span is a term used in a variety of senses. Therefore it is necessary to define it in each case. As in preceding sections of this book, the life span is defined here as the maximal duration of the life of individuals of the given population in actual environmental conditions. I assumed that the maximal age of mollusks, established from sufficient evidence, not only reflects elimination by predators and disease but also mortality due to "old age," i. e., the inner potentialities of the animals under the given environmental conditions. For assessing the maximal life span, it is necessary to indicate what part of the population reaches the given age. Such material is usually lacking, but, whenever possible, the maximal age noted is that reached by about 5–10% of the individuals of the same generation, rather than the age of single long-lived individuals.

Let l_m be the maximal actual individual length in the population and T the age of this individual (i. e., the maximal life span). It follows from equation (76) that

$$T = -\frac{1}{\alpha} \ln\left(1 - \frac{l_m}{l_\infty}\right). \tag{81}$$

For greater convenience we set $\lambda = \frac{l_m}{l_\infty}$.

The quantity λ represents a certain "survival level" which is rather arbitrary because it is in the form of the ratio between the actual maximal length and the theoretical limit reached only after an infinitely long period. As noted previously, however, l_∞ is one of the two parameters determining the form of the growth curve.

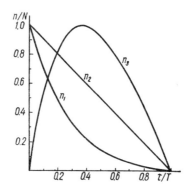

FIGURE 27. Age distribution curves used in the model (explanation in text)

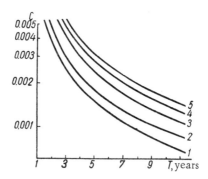

FIGURE 28. Theoretical form of the relationship between the specific production (C) of mollusks and their maximal life span (T):

1) n_3; $\lambda = 0.95$; 2) n_2; $\lambda = 0.95$; 3) n_2; $\lambda = 0.60$; 4) n_1; $\lambda = 0.95$; 5) n_1; $\lambda = 0.60$.

The value of λ for different animal species is totally unknown. If λ for a population in a given biotope is constant, it follows from euqation (81) that the maximal life span (T) is inversely proportional to the growth rate constant (α). On the other hand, a comparison of growth and life span of different mussel species indicates a relationship between λ and α. Mollusks with a greater α possibly have higher values of λ by the end of their life. In this case, according to equation (81) the life span decreases more rapidly with an increase in α than for $\lambda = $ const. However, the available material is not sufficient for a solution of this problem.

Age structure of the population. According to the prevailing view on the age structure, the number of individuals steadily decreases in successive age groups in stationary populations. The data obtained from natural mollusk populations indicate that the very youngest groups usually do not predominate. Most age distribution curves are bell-shaped with the bell situated in the left part of the curve. Clearly, such populations are in a transitional, rather than stationary state. To some extent, the observed condition can also be attributed to a poor catch of small specimens.

For the model three hypothetical types of age structure were chosen, characterized by relatively simple functions and sufficiently representative of most known age compositions of mollusk populations. These curves are shown in Figure 27; they are expressed by the equations

$$n_1 = N\left(1 - \frac{\tau}{T}\right)e^{-b\frac{\tau}{T}}, \tag{82}$$

$$n_2 = N\left(1 - \frac{\tau}{T}\right), \tag{83}$$

$$n_3 = N\left(1 - \frac{\tau}{T}\right)(1 - e^{-b\frac{\tau}{T}}), \tag{84}$$

where $\frac{\tau}{T}$ is the relative age (as a fraction of maximal age), n is the number of individuals of the given relative age, N is total number and b is a constant.

CONSTRUCTION AND ANALYSIS OF THE MODEL

The relationship between specific production, growth rate and life span can be expressed on the basis of the auxiliary equations given above. The specific production, the production and the biomass are given below in accordance with their definitions:

$$C(t) = \frac{P(t)}{B(t)}, \tag{85}$$

$$P(t) = \int_0^T w(\tau, t)\, q_w(\tau, t)\, n(\tau, t)\, d\tau, \tag{86}$$

$$B(t) = \int_0^T n(\tau, t)\, w(\tau, t)\, d\tau. \tag{87}$$

Using equations (76), (79), (80) and (85)–(87), we obtain

$$C = \frac{3\alpha \int_0^T e^{-\alpha\tau}(1 - e^{-\alpha\tau})^2\, n\,d\tau}{\int_0^T (1 - e^{-\alpha\tau})^3\, n\,d\tau}. \tag{88}$$

This completes the model. We shall review the results obtained in solving equation (88) for different values of α and for selected forms of the functions $n(\tau, t)$. Without going into the cumbersome formulas obtained from equation (88) by using n according to equations (82)–(84), we only remark that in all cases C is a function of the ratio $\frac{\alpha}{T}$ $\left(\text{i.e., proportional to } \frac{\alpha}{T} \right)$. The value of the proportionality factor depends on the form of the function n (the solutions were obtained for $\lambda = \text{const.}$).

Since C is inversely proportional to T, the variation of α yields a hyperbola if λ and α are constant.

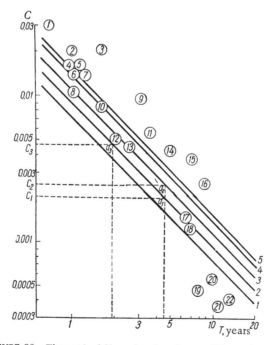

FIGURE 29. Theoretical lines showing the specific production (C) as a function of life span (T) of mollusks with reference to actual data (logarithmic scale; legend as in Figures 28 and 10)

For different values of λ and n, a family of hyperbolas is obtained. Figure 28 shows the curves of C as a function of T for 3 forms of n (equations (82)–(84)) and for α equal to 0.955, 0.85 and 0.6.

It can be seen that for a given value of λ the curve of C is highest at the age structure of type I (equation (82)) and lower at the age structure of type II (equation (84)). This is so because the proportion of young individuals decreases with high q_w values with the transition from a structure of type I to types II and III.

One can also calculate the effect of λ for a given age structure; for a lower value of λ the curve of C will be located higher (lowering of λ leads to a certain "juvenilization" of the population).

Since C is related hyperbolically to T, when plotted on logarithmic coordinates the curve of C as a function of T is a straight line forming an angle of 45° to the axes (Figure 29). Using this graph the relationships between the quantities C, α, T and λ for a constant age structure will be:

1) for λ = const, an increase of α causes an increase of C (in Figure 29, $\alpha_3 > \alpha_1$ and $C_3 > C_1$); 2) for T = const, an increase in λ causes an increase of α and decrease of C (in Figure 29, $\lambda_2 < \lambda_1$ and $C_2 > C_1$.

In selecting mathematical expressions for describing growth, age structure and life span I tried whenever possible to begin with factual data. Now I shall test the theoretical relation between C and T derived from the model against factual data. The circled numbers in Figure 29 indicate values of C vs. T found for 21 species of mollusks (Zaika, 1970 b; see Figure 10). It can be seen that the relative positions and heights of the points are close to the area of the theoretical curves. Consequently, the model accurately describes actual data and explains the behavior of C as a function of T. Of course, values of C for different species can hardly be expected to lie along a single curve or within the band formed by the theoretical curves, since in nature great variations in age structures are encountered and the limits of λ may possibly be wider than those used.

4. LIMITS OF SPECIFIC PRODUCTION OF DIFFERENT ANIMALS AS A FUNCTION OF THE LIFE SPAN

Analysis of the model elucidating the empirically established relationship between specific production and maximal life span in mollusks showed that C is related hyperbolically to T if all other conditions are equal. Moreover, the formula $C = \dfrac{\ln 2}{g}$, used in calculating specific production in microorganisms, also shows a

TABLE 5. Correlation between diurnal specific production and life span of some aquatic animals

Species	c	T	
Rotifers and worms			
Asplanchna priodonta	0.5	6.5	days
Brachionus rubens	1.5	5	"
Dactylogyrus vastator	0.2—0.3	20—25	"
Gyrodactylus elegans	0.28	13—14	"
Limnodrilus newaensis	0.01—0.04	1.5	years
Harmathoe imbricata	0.005—0.001	4	"
Crustaceans			
Daphnia pulex	0.21—0.45	20—30	days
D.longispina	0.12	20—25	"
Bosmina coregoni	0.10—0.15	27—36	"
B.longirostris	0.14—0.15	20—25	"
Chydorus sphaericus	0.13—0.20	22	"
Moina rectirostris	0.25	15	"
Penilia avirostris	0.19	20—25	"
Cyclops sp.	0.12	25—30	"
C.kolensis	0.023	300	"
Epischura baicalensis	0.022—0.031	200—250	"
Calanus finmarchicus	0.019	300—360	"
Limnocalanus johanseni	0.019	180	"
Gammarus lacustris	0.0082	1	year
Gammarus lacustris	0.0055	2	years
G.locusta	0.017—0.048	250—300	days
Gammarelus carinatus	0.008	370—400	"
Dexamine spinosa	0.017—0.020	1	year
Amphithoe vaillanti	0.024—0.050	300	days
Pontoporeia affinis	0.0094	400—450	"
Pontoporeia affinis	0.0052	800—850	"
Micruropus kluki	0.0068	1	year
M.possolskii	0.010	1	"
Gmelinoides fasciatus	0.0080	1	"
Gmelinoides fasciatus	0.0044	2	years
Pontogammarus robustoides	0.018—0.020	400—450	days
Acanthogammarus grewingki	0.0004	10	years
Balanus balanoides	0.0022	8	"
Pandalopsis dispar	0.0020	3	"
Pandalus jordani	0.0013	3	"
P.borealis	0.0010	3	"
P.platyceros	0.0017	4	"
P.hypsinotus	0.0014	3	"
P.danae	0.0020	2.5	"
Mollusks			
Mytilus galloprovincialis	0.0025	7—9	years
Acmaea testudinalis	0.0011—0.0013	6—7	"

FIGURE 5. Contd.

Species	C	T
Acmaea digitalis	0.0035	6—8 years
Mytilaster lineatus	0.0095	3 "
Cardium edule	0.0041	5 "
Abra ovata	0.0049	3—4 "
Rissoa splendida	0.011	1 year
Unio pictirum	0.00035	13 years
U.tumidus	0.00035	11 "
Anodonta anatina	0.00054	10 "
Dreissena polymorphia	0.0014	6 "
Sphaerium corneum	0.0044	300—350 days
Bithynia tentaculata	0.0049	2 years
Adacna vitrea	0.016	1 year
Lacuna pallidula	0.03	300—350 days
Margarita helicina	0.008—0.020	500—600 "
Spisula elliptica	0.020	300—350 "
Psephidia lordi	0.003—0.009	1.5 years
Echinoderms		
Amphiodia urtica	0.0017—0.0022	5 years
Cucumaria elongata	0.0007	10 "
Fishes		
Azov kilka	0.0063	4 years
Anchovy	0.0037	5 "
Baltic lake smelt	0.0039	4—5 "
Round goby	0.0080	4—5 "
Herring	0.0043	5—7 "
Cisco	0.0021	11 "
Omul (Coregonus autumnalis)	0.0040	11—14 "
Bream	0.0071	6—8 "
Sterlet	0.0022	20—25 "

hyperbolic relationship between C and T since the generation time (g) is equal to the life span in microorganisms.

This leads to the assumption that we are dealing with a general law applicable to all organisms. To test this hypothesis, the data on specific production and maximal life span of aquatic animals belonging to different systematic groups were examined. As shown in Figure 30, the points obtained lie within a strip generally corresponding to the assumed hyperbolic relationship between C and T. The experimental data given in Figure 30 and Table 5 represent a

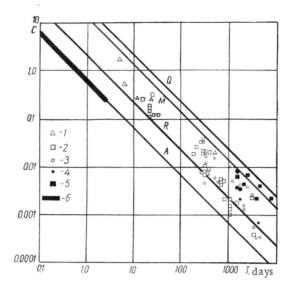

FIGURE 30. Possible range of specific production (*C*) of different
animals as a function of their maximal life span (*T*):

1) rotifers; worms; 2) crustaceans; 3) mollusks; 4) echinoderms;
5) fishes; 6) band indicating the commonest values of *c* and *T*
for infusorians.

sampling of materials presented in Chapter V. The thick line in
Figure 30 represents the commonest values of *C* for bacteria and
infusorians given in Chapter IV. Since the value of *T* is not known
for many species, only a small fraction of the data could be used.
In addition to material on aquatic invertebrates, some data on fishes
are also used (Greze, 1965 b). In evaluating *T* for large animals
(mollusks, fishes) I tried to use the age reached by about 5—10% of
the individuals of the same generation. All I had at my disposal for
small zooplanktonic organisms was scant information on maximal
or average life span, often determined in laboratory conditions.
    In many cases the values not only of *T*, but also of *C*, were deter-
mined approximately or for brief periods. For this reason the comparison involves diurnal annual averages of *C* (for long-lived
mollusks and fishes) as well as average diurnal values of *C* for
relatively short periods (which for small planktonic animals is
usually the period of intensive development). I believe, at any rate,
that the average values of *C* for the period of greatest development
in the biotope are needed in comparing seasonal organisms.
    Although certain points require verification and better definition, the
general trend of the distribution of the points in the graph should
be noted.

The strip in which actual points are distributed in Figure 30 is fairly wide and will probably become even wider with the accumulation of new data. This seems to rule out any hyperbolic relationship between C and T. However, several theoretical considerations seem to deny this assumption. On the basis of different premises, it is possible to estimate both limits of the possible variation of C in animals for any value of T. These limits are such that the width of the strip of possible values of C (for different T) obtained theoretically coincides roughly with the width of the strip including factual points used here.

Models based on different premises will be described below in succession. The numerical evaluation of the limits of C was done with maximum care, so that the limits of C established theoretically are probably wider than the actual limits (at a given T). This probably applies to organisms generally and certainly to some groups having a uniform growth rate and other characteristics limiting variations in different biological parameters, including C.

MODEL BASED ON EXPONENTIAL GROWTH OF THE INDIVIDUAL

This model is naturally crude because exponential growth is not characteristic for organisms. However, the simplicity and clarity of this model allow introduction of corrections, required because the actual growth differs from the exponential type.

In exponential growth, the specific production (C) of the population equals the specific rate (q_w) of weight increase, which is constant and equal in all the individuals in the population. Consequently, the age structure of the population has no significance. This model can be applied to all organisms, especially microorganisms. Simple binary fission in microorganisms doubles weight during the life span. During their life span most multicellular organisms multiply their initial weight many times. In general form, the terminal weight of an animal (w_m, a weight corresponding to the life span T) can be expressed as the product of the initial weight (w_0) and a factor (n) by which weight increased during life:

$$w_m = n w_0. \tag{89}$$

Since in exponential growth

$$w_m = w_0 e^{CT}, \tag{90}$$

equations (89) and (90) yield the following expression for specific production:

$$C = \frac{\ln n}{T}. \qquad (91)$$

According to equation (90), C and T are related hyperbolically for any value of n. Using different values of n, a family of curves which differ from one another by $\ln n$ is obtained. The C—T curve for microorganisms (for $n = 2$) probably represents an aggregate of points, each the lower limit of C for a given T, since no organisms are known in which n is less than 2 (if such did exist, the case of $n < 2$ would be atypical).

Thus, the lower limits of C were established for each T, and the corresponding curve A is shown in the graph (see Figure 30). The left branch of this curve indicates the actual values of C for micro-organisms; the region of values of T (consequently, also of C) most typical of bacterial unicellular algae and infusorians, is delineated. The right branch of the curve should be regarded merely as a theoret-ical limit, since very large values of T (around 100 days or longer) are not typical of microorganisms.

Now let us attempt to use this model to estimate the upper limits of C. For this, it is obviously necessary to estimate the maximal possible values of n. It can be assumed that the highest values of n are characteristic of animals which produce eggs (fishes, mollusks). Some of these values of n are known: Black Sea turbot (Scophthal-mus maeoticus) $n = 2 \cdot 10^7$; Siberian great sturgeon (Huso dauricus) $n = 3 \cdot 10^7$; bluefin tuna (Thunnus thunnus) $n = 1 \cdot 10^8$; Black Sea mussel (Mytilus galloprovincialis) $n = 4 \cdot 10^8$. These values were calculated as the ratio between maximal individual weight and weight of the egg from initial data found in various reviews and textbooks. There also may be values of n somewhat higher than those indicated. It is assumed that the highest possible values of n approach $1 \cdot 10^9$, which yields the upper limit of C for each T according to equation (91).

The curve Q corresponding to this is given in Figure 30. The actual data evidently lie within the strip defined by curves A and Q. Moreover, most points lie above curve R corresponding to $n = 10$; high values for n are more probable for high T and low n values (corresponding to low values of C) for low T. This is based on general considerations, namely, that a considerable increase in weight during the life span clearly requires more time, if other conditions are equal.

MODEL BASED ON S-TYPE GROWTH

Of course, special models of this category were used in the preceding section in analyzing productivity of mollusks. All the curves in Figure 29 lie between curves R and M (see Figure 30) in a region occupying the central part of the strip based on the exponential model. The lines shown in Figure 29 relate to special cases and do not permit calculation of the limits.

The model, based on an S-shaped type of individual growth, does not permit determination of the lower limit of values of C for each T. For the S-shaped type of growth, adults usually have low specific increments (even when production of gonadal products is taken into account). Accordingly, the lower limit of C is determined here from fluctuations in the age structure of the population. Field observations show that the entire population of many species may be composed entirely of adults at a given moment; therefore, regardless of T, the theoretical lower limit of C in S-type growth is zero. However, such a situation is not typical, since the population cannot always consist only of adults. Hence, it is necessary to base the calculation on age structure averaged over a certain period. This period apparently should not be less than T. The average value of C for this period would obviously be greater than zero. Unfortunately, we know very little of the demography of populations and it is impossible to define quantitatively "normal limits" below which C cannot occur in S-type growth.

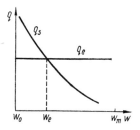

FIGURE 31. Diagram illustrating the decrease of the specific growth rate in S-type growth (q_s) and its constancy in an exponential growth (q_e)

The evaluation of the upper limit of C is possible within the framework of the model. From equation (78) for animals with S-type growth it follows that $q_l = \alpha$ as $\tau \to 0$ (after the module). This means that the parameter α reflects the specific rate of linear growth at $\tau = 0$. According to equation (80), $q_w = 3q_l$; therefore, at $\tau = 0$ we obtain

$$q_w = 3\alpha. \qquad (92)$$

The condition $\tau = 0$ means that $w = 0$. Since growth actually
begins from some $w_0 > 0$ and the value of q_w in an S-type growth
decreases with age, it follows that equation (92) reflects a practically inaccessible limit of q_w. To obtain the upper limit of C from
the limit value $q_w = 3\alpha$, let us consider the extreme case in which the
population consists (theoretically) entirely of individuals weighing
$w = 0$. In such a case the population has the maximal possible
specific production:

$$C = 3\alpha. \tag{93}$$

To calculate α for each T according to equation (81) it is necessary to
assume a given value of λ, which increases as the parameter α
increases for a given T. As a precaution a high value of λ, namely
$\lambda = 0.99$, has deliberately been chosen. Then $\alpha = \dfrac{4.6}{T}$ from equation
(81) and $C = \dfrac{13.8}{T}$ from equation (93). This is the upper limit of C
for each T according to this version of the model. The corresponding curve M is indicated in the graph. Almost all the actual points
(except 3 points for fishes) lie below this line. It should be noted
that on the basis of the preceding model (exponential growth),
$C = \dfrac{13.8}{T}$, i.e., curve M, is also obtained for $n = 10^6$.

To continue the comparison of the two models, Figure 31 shows
the ratio of the specific rates of individual growth, namely, q_e for
exponential growth and q_s for S-type growth for the same w_0, w_m and T.
The conversion from q_e and q_s, respectively, to C_e and C_s yields
either $C_e < C_s$ or $C_e > C_s$, depending on the age structure of the
population. $C_e = C_s$ at the intersection of the curves in Figure 31,
which is obtained for a certain w_e. If individual weight less than w_e
predominates in the population (in biomass), it follows that $C_s > C_e$.

The curve in Figure 32 shows w_e as a fraction of w_m in relation
to the value of n. (The curve is based on the assumption that
$\dfrac{w_m}{w_\infty} = 0.9$ in S-type growth.) Obviously, the probability of $C_s > C_e$
decreases with the increase in n. Indeed, $C_s > C_e$ for $n = 10^6$ only
if the bulk of the biomass consists of individuals weighing less than
$0.1\, w_m$, which is hardly a typical situation. Accordingly, $C_s < C_e$ is
hardly probable if $n \leqslant 10$.

Thus, in the transition from the model of exponential growth to
a more real model (S-type growth), we see that with S-type growth
and for $n = 10$ it is hardly probable that C will be located below the
left branch of curve R. This can happen in the right branch of curve R
but for a large T low n values are quite improbable. Similar

considerations indicated that in S-type growth, C must not be above curve M; indeed, this curve was obtained as an upper limit directly from the model of S-type growth.

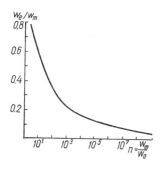

FIGURE 32. Curve showing change in the ratio between weight (w_e) at $C_s = C_e$ and the maximal individual weight (w_m) as a function of the ratio between the maximal weight and the minimal weight (n)

MODEL BASED ON CONCEPTS OF THE NATURAL RATE OF POPULATION GROWTH

In Chapter III it was shown that the factor of natural increase of the population (r) can serve as a minimal estimate of specific production. Several researchers outside the Soviet Union have examined the problem of the natural increase of populations. These publications are outside the scope of this book and will not be discussed here. Moreover, this trend in ecology has been treated in a number of reviews (Slobodkin, 1962; MacFadyen, 1965). Briefly, the problem can be stated as follows. Let us assume that the size of the population grows exponentially:

$$N_t = N_0 e^{rt},$$

where r is the factor of natural increase. The problem is to determine the value of r in different populations, which is a rather difficult task involving preparation of "life tables" of the population based on prolonged and detailed observations.

These works also employ the indexes R (rate of population replacement) and T' (length of the generation). If $R_0 = 1$ the population is replaced during the period T' for which $r = 0$, i.e., the population does not increase. For a growing population we have

$$R_0 = e^{rT'},$$

(94)

hence

$$r = \frac{\ln R_0}{T'} . \qquad (95)$$

The replacement rate R_0 indicates the factor by which population size increases in time T', and so equation (95) is clearly quite similar to (91). The difference is that the former refers to exponential increase in the number of individuals rather than in their weight. As already noted (see Chapters I and III), such an approach is possible for determination of productivity, although authors of works on natural increase in populations do not mention productivity. Replacing population size by biomass, we have $R_0 \approx n$ and $T' \approx T$ (for multicellular animals the values of T' and T are determined approximately, i. e., practically they are difficult to distinguish).

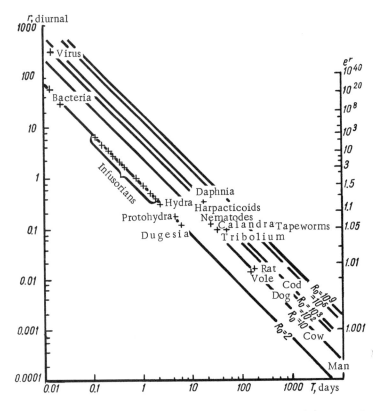

FIGURE 33. Relationship between rate of population increase (r), generation time (T) and the index $R_0 = e^{rT}$ (after Fenchel, 1968); $r = \dfrac{\ln R_0}{R_0 e^{rT}}$

Finally, it can be demonstrated that $C \approx r$ if the quantity of mortality is disregarded. In Chapter I we introduced the equation $r = b - m$ where b is the reproduction factor and m the elimination factor, as well as $C = b$. Hence $C = r + m$, and if $r \gg m$ we have $C \approx r$.

Thus, the value of r is only a rough estimate of C, but this model should be mentioned despite the fact that it shows the limits of C no better than the first model. In this case it is noteworthy that if results are put in comparable form seemingly remote fields of population ecology yield identical conclusions. Perusal of the graph shows that it is practically identical with our graphic analysis of the relationship between C and T. This graph was first published by Smith (1954) and later appeared in a number of reviews. Figure 33 is taken from Fenchel (1968) who added data on infusorians to those of Smith.

The models described above lead to the conclusion that the curves A and Q provide possible limits of specific production for different values of T; moreover, the specific production of many or most species lies between the curves R and M. The extreme values of C (along the curves A and Q) and the average value $\bar{C}$ expected at a given T are indicated below:

T, days	5	15	30	60	100
C, limits	0.15–4.0	0.05–1.4	0.025–0.6	0.01–0.3	0.007–0.2
$\bar{C}$	0.8	0.3	0.15	0.07	0.04

T, years	1	3	5	10
C, limits	0.002–0.05	0.0006–0.02	0.0004–0.01	0.0002–0.0001
$\bar{C}$	0.01	0.003	0.002	0.001

For microorganisms, which reproduce by binary fission, each T corresponds to a given value of C, in particular:

T, hours	1	2	4	8	12	20	40	50	80
C, (diurnal)	16.6	8.3	4.1	2.1	1.4	0.83	0.41	0.31	0.22

In characterizing the results of the above analysis of specific production as a function of maximal life span, the following two points should be noted.

1. From a theoretical viewpoint it is important that results comparatively close to the limits of C are obtained from different premises, i. e., from different models of the relationship between C and T. This both confirms the general conclusions and shows that ecological and, in part, physiological concepts can yield consistent solutions to a given problem.

2. Analysis of the limits of C and determination of average values of C expected for each T allow an estimate of the specific production of the population based only on the maximal life span. Naturally, this does not relieve us of the necessity to study further the values of specific production of different populations, but it considerably simplifies making a rough estimate of productivity where needed.

5. RELATIONSHIP BETWEEN SPECIFIC PRODUCTION, POPULATION BIOMASS AND SIZE OF ANIMALS

The main results obtained here are based on the study of values of specific production of the population. I stressed that specific production is the best comparative index and is essential for studying many problems, although the value of production of the population according to biotope area or volume units is required in some cases. I emphasized the analysis of specific production because it seems to me to be essential at the present stage of productivity research. Indeed, hydrobiologists have accumulated considerable material on biomasses of different species and groups in a variety of biotopes. With the development of a reliable and rapid method for determining specific production of the most typical and abundant species of the main systematic groups, all the available data on biomass can be used for calculating production.

The parallel study of biomass and specific production will probably reveal a number of theoretically interesting principles which now can only be guessed. This section deals with the most hypothetical of these principles. These have not yet been specially studied, but deserve attention in future experimentation.

Prevailing concepts on the relationship between production and biomass are based on somewhat vague experimental and field observations and relate not only to populations but also to various superpopulational systems. On the basis of considerable data from culture of food invertebrates and fishes, Shpet (1968) concluded that large animals reach smaller numbers but greater biomass than small animals reared in identical tanks and under optimal food conditions. The same trend is observed in many species under natural conditions. Shpet attributes this phenomenon to the higher metabolic rate of the smaller forms, which require more "living space" and, especially, more food per unit biomass.

These data must be analyzed in detail and studied further. What is the present theoretical basis for the relationship described by Shpet? First, there is the link between the ration and body size.

Several publications have shown that the ratio between the ration and body size in various animal species decreases during growth. Recent works on different crustacean species have established a quantitative relation between ration and body weight (Inoue, 1964; Sushchenya and Khmeleva, 1967; Abolmasova, 1969). Sushchenya and Khmeleva (1967) found that the specific ration is related to body size in different crustacean species like that observed during the growth of an individual. For the species examined, the following general equation was obtained (at 20°C):

$$R = 0.0746w^{0.80}. \tag{96}$$

For two species of different size $(w_1 > w_2)$ it can safely be assumed that $\frac{R_1}{w_1} < \frac{R_2}{w_2}$.

What is the proportion between the biomasses of the compared species? It can be shown for the case in question that the larger animals have a greater biomass $(B_1 > B_2)$. If the animals are grown in tanks containing equal amounts of food, then when the biomass limits are reached food will become a limiting factor and its total consumption may be nearly equal in each species:

$$\frac{R_1}{w_1} B_1 = \frac{R_2}{w_2} B_2. \tag{97}$$

This situation may be realized in the case where species having similar behavior with respect to the offered food (i.e., species having a similar food spectrum) are compared for $\frac{R_1}{w_1} < \frac{R_2}{w_2}$; however, it follows from equation (97) that $B_1 > B_2$.

Assuming that the compared species have the same factor of food assimilation, this means that in the case described the species assimilate the same amount of food A. We know (see Chapter I) that the production of a population can be expressed by the equations

$$P = A - T,$$
$$P = BC,$$

where A is assimilation, T is losses in metabolism and C is specific production. It follows that

$$A = BC + T,$$

and if $A_1 = A_2$, then

$$B_1 C_1 + T_1 = B_2 C_2 + T_2. \tag{98}$$

If the two species have the same specific production $(C_1 = C_2)$, despite their different size, the biomass of the larger animals must be greater since the smaller animals have greater specific losses and for $C_1 = C_2$ and $B_1 \leqslant B_2$, the congruency of equation (98) is disrupted.

It can be assumed that the larger animals have a lower specific production $(C_1 < C_2)$. In particular, if the large animals in this example live longer they will have a lower specific production (see Chapter VI). If it is also assumed that $B_1 \leqslant B_2$, then equation (98) is violated since it is impossible that $T_1 > T_2$. Thus, for $C_1 < C_2$, the larger animals also have a greater biomass.

Since we are concerned mainly with the relation between biomass and specific production it would be desirable to examine some implications of the simple model described above without reference to data on nutrition or respiration. At the established level of utilization of the food supply, each animal species shows a given level of production equal to the difference between assimilation and losses. When the level of the food supply limits production of the consumer, the latter (theoretically) adapts to a given production level by changing either the biomass or the specific production, since $P = BC$. Indeed, the biomass and specific production are generally inversely proportional.

A trophic level composed of a number of species populations possesses a mechanism varying the average specific production without affecting the specific production of the different species. This mechanism involves changes in the proportions of the biomasses of species having different specific productions. For example, in succession there may be an initial peak of small organisms with high C values followed by an increase in the proportion of long-lived forms with low C values. As a result, the total biomass of the trophic level increases with a concomitant decrease in the average specific production. It is stressed that this can occur with a constant food supply.

Turning to separate species, the specific production decreases with a poor food supply. There is a decrease in the growth and reproduction rates of the individuals. The resulting decrease of the proportion of juveniles in the population leads to a lowering of C. For "normal" or "adequate" food supply, however, the mechanism controlling the value of production can hardly be expected to lower specific production. For example, where animals are placed in tanks containing a given food concentration the population grows rapidly at the highest values of C the species can reach in the given conditions. Production increases along with the biomass. The biomass stops growing when food becomes the limiting factor.

Under such circumstances the biomass may temporarily exceed the optimal level, causing some decrease of C. However, the general result, according to this interpretation, is that biomass growth ceases while C remains high.

Thus, I assume that the population tends to retain the highest possible specific production for given conditions, whereas the biomass remains at the level limited by the given specific production and food supply. This assumption needs verification. We can only refer to data indicating that the specific production of the population is far less variable than the biomass. It was found, for example, that the biomass of planktonic crustaceans varies tens of times while specific production varies by a factor of only 1.2–2 (Zaika and Malovitskaya, 1967). The following table shows average annual values of B, P and P/B (annual specific production) for three zooplankton species in the Sea of Azov (Table 6).

TABLE 6. Variability of biomass (B, mg/m^3), production (P, mg/m^3) and specific production (P/B) of three populations in the Sea of Azov in 1963–1965

Acartia clausi			Calanipeda aquae-dulcis			Centropages kröyeri		
B	P	P/B	B	P	P/B	B	P	P/B
4	54	14.1	22	877	40.2	1.3	30.4	23.4
18	299	16.3	42	1389	32.6	5.8	71.7	12.3
39	631	16.2	55	2055	37.0	7.6	192.4	25.3
55	936	17.0	374	9898	26.5	10.9	182.6	16.7
214	3031	14.2	575	17953	31.2	40.9	671.8	16.4
219	3676	16.8	686	14696	21.4	88.1	985.1	11.2

Ratio between maximal and minimal values

55	63	1.2	31	20	1.9	70	33	2.0

In Table 6, the values of B are arranged in ascending order from top to bottom. In the bottom row of the table the ratios between maximal and minimal values are given. It can be seen that the biomass is far more variable than the specific production. On the basis of long-term studies of productivity of copepods in the Black Sea, Greze et al. (1968) similarly regard specific production as relatively constant compared to biomass.

It should be noted that the possible effect of feeding conditions on the individual growth rate was not taken into account in calculating P and C by the workers cited above. However, it is hard to believe that under the influence of changes in the food supply the specific production of these populations could have changed ten-fold during the years studied.

Thus, invertebrates react to a change of environmental conditions mainly by a relatively large fluctuation in biomass together with minor changes in specific production. In microorganisms, on the other hand, the mechanism controlling biomass increase and specific production is such that these values change more or less together (by varying the rate of cell division). For this reason the biomass increase decelerates together with a decrease in specific production when the possible feeding limit is reached (naturally, this applies only to cases in which feeding does not control the biomass of the microorganisms).

A similar phenomenon (i. e., a marked change in specific production) can occur in other organisms if the population has the simplest structure, i. e., consists entirely of organisms of the same age. Brocksen et al. (1968) present data on the development of young sockeye salmon in lakes; with food as the limiting factor, the rate of weight increase becomes lower as biomass increases. Such populations are obviously in a peculiar position, since they cannot regulate their biomass increase by altering the reproduction rate (since the population consists entirely of young forms). The growing specimens exhaust the food supply, resulting in a decrease in growth rate and specific production.

The pressure of predators has great significance in regulating the biomass. In such cases where predators suppress biomass increase, the limiting significance of food supply is diminished. In other words, factors which decrease specific production are inhibited. The pressure of predators thus maintains a high specific production in the population.

CONCLUSION

Research into problems of productivity has a long history and covers a wide range of expanding fields in hydrobiology. Productivity is studied in systems of different complexity: species populations, communities (phytoplankton, bacterioplankton, zooplankton, benthos), biocenoses and ecosystems. Work on production of terrestrial organisms has followed similar lines.

The systems enumerated are hierarchically subordinated to one another: a biocenosis consists of interdependent communities, which in turn are composed of species populations. However, there is no strict order in studies of productivity of systems of different levels of complexity, i. e., productivity on the community level is studied together with the productivity of populations, and not after the latter is studied. At the same time, it is evident that the state of knowledge on productivity of populations directly reflects on the standard of concepts on productivity in more complex systems. In turn, progress in the field of population productivity closely depends on research into such fundamental parameters as individual growth, fecundity, age structure of the population, etc.

The problem of productivity can therefore be subdivided according to **objects studied,** with both formulation and treatment of the problem depending on the complexity of the biological system as evidenced by the specific qualitative features of the system. At the same time the study of the productivity in different systems involves a series of identical stages including logical analysis of the general concepts, selection of quantitative indexes of productivity, working out of procedures, performance of experimental or field research, analysis of results, detection and interpretation of relationships between productivity and other factors and, finally, practical application of information obtained. The existence of these **stages of research** allows subdivision of the problem on a new plane. It is easily seen that the sequence of these stages is more logical than temporal, since no single stage belongs entirely to the past or future. However, the basic types of systems differ profoundly in certain aspects. There are many rough estimates of productivity of large communities and, at the same time, rather accurate determinations of production indexes in a number of species populations.

Initially, productivity was measured by the general level of vital development in the biotope or by semi-intuitive estimate of the rate of turnover of organic matter. The trend toward accurate calculation of concretely defined indexes became evident later. Indexes showing the general level of development (density, biomass, etc.) are now seldom used for characterizing the productivity of a system; they serve instead as preliminary data for determining production indexes in the strict sense (production, specific production, gross production, potential production, etc.).

The accumulation of material on the productivity of different populations created conditions necessary for detecting and interpreting quantitative relationships between production indexes and the "fundamental parameters" of the population.

Research on animal productivity is now conducted mainly on the specific population level, and, to a lesser extent, on the community level. It is possible to evaluate production in a variety of ways differing from one another in the kind of initial biological data used. The method of choice for determination of the production is that for which the most reliable initial data are available. Regardless of the method selected, however, production can always be presented as the product of specific production and the biomass of the system. Since most works in hydrobiology contain data on biomass, these can be used in production determination if specific production is known. Because of this, studies of specific production are of paramount importance at present.

Specific production is a far more stable index than biomass of a species population; it includes the specific weight increase rates of the individuals. Research on specific production is closely associated with studies of animal growth and population structure. I have indicated some trends of comparative studies of the weight increase rate and specific production of the population. General quantitative patterns linking size and growth and production capacities of the organisms have been discovered. A relationship has also been established between specific production and the life span of the individuals. More thorough and accurate future studies should reveal various aspects of the relationships between the specific production and basic morphophysiological indexes in animals. Still another interesting trend for future research is that under equal conditions biomass and specific production are not totally independent.

It can be stated in conclusion that populations tend to maintain the highest possible specific production under existing conditions, whereas the biomass remains at the level dictated by the given combination of specific production and food supply. Hence the relationship between specific production and biomass with respect to food supply should be investigated.

In the present work, the productivity of populations was evaluated preferably for natural conditions. Turning to commercial conditions, productivity poses a number of problems going beyond the framework of ecology. Among these, two problems deserve particular attention: 1) to what extent can the specific production be increased by artificial changes of environmental conditions and by controlled breeding; 2) what is the optimal strategy in selecting animals for prey and for breeding in terms of the relationship between their size, biomass and specific production. These topics are actually not new, but are of great theoretical interest in a number of ways.

Agricultural practice demonstrated long ago that the yield and commercial productivity of plants and animals can be markedly improved by favorable conditions and selection. We now turn to the question of the ratio between the productivity level reached in animal breeding and the theoretical limits of possible growth rate and specific production, which was discussed in this work. Since the theoretical upper limits of the specific growth rate (and specific production) of organisms of different sizes and life span were obtained mainly from material on natural populations, it would be desirable to determine the influence of man on these relationships. It is quite obvious that intentional changes in the age structure of the stock and improvements in culture conditions raise specific production much closer to the potential limit for the given species. To exceed the theoretical limit established for natural populations, however, the physiological properties (growth rate for multicellular organisms and division rate for unicellular ones) have to be altered radically. Special analysis of such data in comparison with material on natural populations will determine how much man has changed the nature of cultivated organisms and whether natural potentials of growth and specific production in these species have been exceeded.

Apparently, the greatest advances in this direction have been made in the culture of microorganisms.

The second problem mentioned is also not new, but is becoming increasingly important because of the expansion of the fisheries and cultivation of unusual species in the search for new food and raw material resources.

The selection of new items for fisheries and cultivation must be guided not only by criteria such as biochemical composition, palatability, ease of processing, etc., but also by the relationship between specific production of the population and individual size and biomass reached per unit of biotope volume or area. It is obviously necessary to determine the optimal combination of such parameters, since they may vary in opposite directions in some species.

Finally, the use of the total amount of weight increase in pro-duction research is justified only at initial stages as part of an extensive comparative study. A more profound analysis of pro-ductivity of commercial and valuable species requires numerous additional indexes characterizing the production of the most important items such as meat, milk, fur, etc.

Applied production research is obviously promising. However, this does not detract from the importance of the general theory of productivity which must be developed in close association with various fields of animal ecology and physiology.

REFERENCES

Abolmasova, G.I. Relationship between Food Ration and Body Weight in Higher Crustaceans. – In: Voprosy morskoi biologii. Kiev, "Naukova Dumka." 1969.

Adolph, E.F. The Regulation of Adult Body Size in the Protozoan Colpoda.– J.Exp.Zool., 53:2. 1929.

Afanas'eva, E.L. Comparison of Productions Determined by Different Methods of Epischura baicalensis Sars from Lake Baikal. – In: Metody opredeleniya produktsii vodnykh zhivotnykh. Minsk, "Vysheishaya Shkola.". 1958.

Alm, G. Die quantitative Untersuchung der Bodenfauna und Flora in ihrer Bedeutung für die theoretische und angewandte Limnologie. – Verh.int. Verein.theor.angew.Limnol. 1924.

Arabina, I.P. Seasonal and Annual Dynamics and Production of Zoobenthos in Lakes Narochi, Myastro and Batorin. Author's Summary of Candidate Thesis. Minsk. 1968.

Baitsell, G.A. Experiments on the Reproduction of the Hypotrichous Infusoria. – J.Exp.Zool., 13:1. 1912.

Baitsell, G.A. Experiments on the Reproduction of the Hypotrichous Infusoria. 2. – J.Exp.Zool., 16:2. 1914.

Bary, B.M. Notes on Ecology, Distribution and Systematics of Pelagic Tunicata from New Zealand. – Pacif.Sci., 14:2. 1960.

Bazikalova, A.Ya. Some Data on the Biology of Acanthogammarus grewingki (Dub).– Trudy Baikal.Limnol.Sta., 14. 1954.

Beers, C.D. On the Possibility of Indefinite Reproduction in the Ciliate Didinium without Conjugation or Endomixes. – Am.Nat., 63:125. 1929.

Beers, J.R. and G.L. Stewart. Microzooplankton in the Euphotic Zone at Five Locations Across the California Current. – J.Fish.Res.Bd.Can., 24:10. 1967.

Bekman, M.Yu. Biology of Gammarus lacustris Sars from Water Bodies in the Baikal Area. – Trudy Baikal.Limnol.Sta., 17. 1954.

Bekman, M.Yu. Some Features of the Distribution and Production of Mass Zoobenthic Species in the Maloe More. – Trudy Baikal.Limnol.Sta., 17. 1959.

Bekman, M.Yu. Ecology and Production of Micruropus possolskii Sow. and Gmelinoides fasciatus Stebb. – Trudy Limnol.Inst. Sibirsk. Otdel. Akad. Nauk SSSR, 2, No.22:1. 1962.

Bekman, M.Yu. and V.V. Menshutkin. Analysis of the Production Process in Populations of Very Simple Structure. – Zh. Obshch.Biol., 25:3. 1964.

Bertalanffy, L. von. A Quantitative Theory of Organic Growth. – Hum.Biol., 10:2. 1938.

Bogorov, V.G. The Primary Production of the Ocean and its Exploitation. – Vestnik Akad. Nauk SSSR, 9. 1966.

Bogorov, V.G. Productive Areas of the Ocean. – Priroda, 10. 1967a.

Bogorov, V.G. Biological Transformation and Conversion of Energy and Matter in the Ocean. – Okeanologiya, 5. 1967b.

Bogorov, V.G. Problems of Oceanic Productivity. – Gidrobiol. Zh., 3:5. 1967c.

Borror, A.C. Morphology and Ecology of the Benthic Ciliated Protozoa of Alligator Harbor, Florida. – Arch. Protistenk., 106:4. 1963.

Borutskii, E.V. Biomass Dynamics of Chironomus plumosus in the Profundal Part of Lake Beloe. – Trudy Limnol. Sta. Kosino, 22. 1939a.

Borutskii, E.V. General Biomass Dynamics of the Profundal Part of Lake Beloe. – Ibid. 1939b.

Boysen-Jensen, P. Valiation of the Limfjord, 1. – Rep. Dan. Biol. Stn., 26. 1919.

B r e g m a n, Yu.E. Growth and Production of the Rotifer A s p l a n c h n a p r i o d o n t a in the Eutrophic Lake Drivyaty. – In: Metody opredeleniya produktsii vodnykh zhivotnykh. Minsk, "Vysheishaya Shkola." 1968.

B r o c k s e n, R.W., G.E.D a v i s, and C.E.W a r r e n. The Analysis of Trophic Processes on the Basis of Density-Dependent Functions. – Symp. Mar. Food Chains, Denmark, Contrib., No.30. 1968.

B r o t s k a y a, V.A. and L.A.Z e n k e v i c h. Biological Productivity of Marine Waters.– Zool. Zh., 15. 1936.

B u t l e r, T.H. Growth, Reproduction and Distribution of Pandalid Shrimps in British Columbia. – J. Fish. Res. Bd. Can., 21:6. 1964.

B y k h o v s k i i, B.E. Monogenetic Trematodes, their Taxonomy and Phylogeny. – Moscow-Leningrad, Akad. Nauk SSSR. 1957.

C a l k i n s, G.N. U r o l e p t u s m o b i l i s Engels. – J. Exp. Zool., 29:2. 1919.

C h m y r, V.D. Radiocarbon Method for Determination of Production of Zooplankton in Natural Populations. – Dokl. Akad. Nauk SSSR, 173. 1967.

C l a r k e, G.L., W.T.E d m o n d s o n, and W.E.R i c k e r. Mathematical Formulation of Biological Productivity. – Ecol. Monogr., 16, No.4. 1946.

C o h e n, B.M. On the Inheritance of Body Form and Certain Other Characteristics in the Conjugation of E u p l o t e s p a t e l l a. – Genetica, 19:40. 1934.

C o m i t a, G.W. A Study of a Calanoid Copepod Population in an Arctic Lake. – Ecology, 37:3. 1956.

C o o p e r, N.E. Dynamics and Production of a Natural Population of a Freshwater Amphipod, H y a l e l l a s z t e c a. – Ecol. Monogr., 35. 1965.

D a v i s, C.C. On Questions of Production and Productivity in Ecology. – Archiv. Hydrobiol., 59:145. 1963.

D e m o l l, R. Betrachtungen über Produktionsberechnungen. – Archiv Hydrobiol., 18:3. 1927.

D r a b k o v a, V.G. Population Dynamics, Generation Time and Production of Bacteria in Waters of Lake Krasnoe (Punnus Yarvi). – Mikrobiologiya, 34, 6. 1965.

D r a g o l i, A.L. On the Relationships between Varieties of Black Sea Mussel (M y t i l u s g a l l o - p r o v i n c i a l i s Lam.). – In: Raspredelenie donnykh zhivotnykh v yuzhnykh moryakh. Kiev, "Naukova Dumka." 1966.

E d m o n d s o n, W.T. Reproductive Rates of Rotifers in Natural Populations. – Memorie Ist. ital. Idrobiol., 12. 1960.

E g g e r t, M.B. Seasonal Changes in the Infusorian Fauna in the Plankton of the Selenga Area of Baikal. – Gidrobiol. Zh., 3:3. 1967.

E g g e r t, M.B. Ecology and Density of Holotricha and Peritricha in the Selenga Area of Baikal. – Gidrobiol. Zh., 3:3. 1968.

F a u r e - F r e m i e t, E. Growth and Differentiation of the Colonies of Z o o t h a m n i u m a l t e r n a n s (Clap. et Lachm.). – Biol. Bull., 58:28. 1930.

F a u r e - F r e m i e t, E. Le rythme de maree du S t r o m b i d i u m o c u l a t u m Gruber. – Biol. Bull., 82:3. 1948a.

F a u r e - F r e m i e t, E. The Ecology of Some Infusorian Communities of Intertidal Pool. – J. Anim. Ecol., 17:2. 1948b

F e n a u x, R. Ecologie et biologie des Appendiculaires mediterraneans. – Vie Milieu, Suppl., 16. 1963.

F e n c h e l, T. The Ecology of Marine Microbenthos. I. – Ophelia, 4:2. 1967.

F e n c h e l, T. The Ecology of Marine Microbenthos. III. – Ophelia, 5:1. 1968.

F i n e n k o, Z.Z. Primary Production in the Black Sea and Sea of Azov and the Tropical Part of the Atlantic. Candidate Thesis. Minsk. 1965.

F i n e n k o, Z.Z. and V.E.Z a i k a. Particulate Organic Matter and its Role in the Productivity of the Sea. Symp. Mar. Food Chains, Denmark. 1968.

F i s h, J.D. The Biology of C u c u m a r i a e l o n g a t a (Echinodermata: Holoturioidea). – J. Mar. Biol. Ass. U.K., 47:1. 1967.

F o r d, E. On the Growth of Some Lamellibrachs in Relation to Food Supply of Fishes. – J. Mar. Biol. Ass. U.K., 13:3. 1925.

F r a n k, P.W. The Biodemography of an Intertidal Snail Population. Ecology, 46:6. 1965.

G a l k o v s k a y a, G.A. The Productive Capacities of Planktonic Rotifers. – Nauchn. Dokl. Vyssh. Shk., Biol. Nauki, **3**. 1963.

G a l k o v s k a y a, G.A. The Production of Planktonic Rotifers. – In: Metody opredeleniya produktsii vodnykh zhivotnykh. Minsk. 1968.

G a l k o v s k a y a, G.A. and V.P. L y a k h n o v i c h. The Production of Pond Zooplankton. – Gidrobiol. Zh., 2:4. 1966.

G a m b a r y a n, M.E. On a Method for Determining Generation Time of Microorganisms in Bottom Sediments. – Mikrobiologiya, 34:6. 1965.

G a v r i l o v, S.I. Productivity of the Zoobenthos of Some Commercial Lakes in Belorussia. Author's Summary of Candidate Thesis. Minsk. 1969.

G e i n r i k h, A.K. On the Production of Copepods in the Bering Sea. – Dokl. Akad. Nauk SSSR, 111:1. 1956.

G r e e n l e a f, W.E. The Influence of Volume of Culture Medium and Cell Proximity on the Rate of Reproduction of Infusoria. – J. Exp. Zool., 46:2. 1926.

G r e z e, V.N. The Production of P o n t o p o r e i a a f f i n i s and a Method for its Determination. – Trudy Vses. Gidrobiol. Obshch., **3**. 1951.

G r e z e, V.N. Growth Rate and Production Capacities of Fish Population. – Gidrobiol. Zh., 1:2. 1965 b.

G r e z e, V.N. Production Rates in Populations of Pelagic Copepoda in Baikal. – In: Krugovorot veshchestva i energii v ozernykh vodoemakh. Moscow, "Nauka." 1967a.

G r e z e, V.N. Production Rate of Populations of Heterotrophic Marine Organisms. – Vopr. Biookeanogr. Kiev, "Naukova Dumka." 1967b.

G r e z e, V.N. and E.P. B a l d i n a. Population Dynamics and Annual Production of A c a r t i a c l a u s i Giesbr. and C e n t r o p a g e s k r o y e r i C. in the Neritic Zone of the Black Sea. – Trudy Sevastopol. Biol. Sta., **17**. 1964.

G r e z e, V.N., E.M. B a l d i n a, and O.K. B i l e v a. Production of Planktonic Copepods in the Neritic Zone of the Black Sea. – Okeanologiya, 8:6. 1968.

G r e z e I.I. and V.N. G r e z e. Specific Production of Populations and Some Amphipods in the Black Sea. – Zool. Zh., **48**:3. 1969.

G r i g a, R.E. Development of Some Harpacticoida in the Black Sea. – Trudy Sevastopol'. Biol. Sta., **13**. 1960.

G r ø n t v e d, J. Preliminary Report on the Productivity of Microbenthos and Phytoplankton in the Danish Wadden Sea. – Meddr. Danm. Fisk. – Havunders., N.S., 3:12. 1962.

H a l l, D. An Experimental Approach to the Dynamics of a Natural Population of D a p h n i a m e n d o t a e. – Ecology, 45:5. 1964.

H e n s e n, V. Über die Bestimmung des Planktons oder des im Meere treibenden Materials an Pflanzen und Tieren. – V. Bericht d. Komm. z. wiss. Unters. d. Deutschen Meere, **12–16**. Kiel. 1887.

H e t h e r i n g t o n, A. The Constant Culture of S t e n t o r c o e r u l e u s. – Arch. Protistenk., 76:118. 1932.

H e t h e r i n g t o n, A. The Role of Bacteria in the Growth of C o l p i d i u m c o l p o d a. – Physiol. Zool., 7:4. 1934.

H e t h e r i n g t o n, A. The Precise Control of Growth in a Pure Culture of a Ciliate G l a u c o m a p y r i f o r m i s. – Biol. Bull., 70:3. 1936.

I e r u s a l i m s k i i, N.D. Nitrogen and Vitamin Nutrition of Microbes. – Moscow-Leningrad, Akad. Nauk SSSR. 1949.

I e r u s a l i m s k i i, N.D. Determination of the Growth Rate of Aquatic Microorganisms on Periphyton Slides. – Mikrobiologiya, 33:5. 1954.

I e r u s a l i m s k i i, N.D. Foundations of Microbial Physiology. – Moscow, Akad. Nauk SSSR. 1963.

I n o u e, M. On the Amount of Food Required by the Japanese Spiny Lobster P a n u l i r u s j a p o n i c u s (V. Siebold) Kept in Cage in Relation to Size and Temperature. – Bull. Jap. Soc. Scient. Fish., 30:5. 1964.

Ioffe, Ts.I. and L.P. Maksimova. Biology of Some Crustaceans Suitable for Acclimatization in Artificial Reservoirs. − Izv. Gos. Nauchno-Issled. Inst. Ozern. Rechn. Ryb. Khoz., (GosNIORKh), 67. 1968.

Ivanov, A.I. Distribution and Resources of Mussels in the Black Sea. − In: Biologiya, tekhnika promysla i pererabotka midii. − Vses. Nauchno-Issled. Inst. Morsk. Ryb. Khoz. Okeanogr. (VNIRO), Ob"ed. Nauchno-Tekh. Izdat. (ONTI), Moscow. 1965.

Ivanov, A.I. Growth of the Black Sea Mussel (Mytilus galloprovincialis Lam.) on the Odessa Bank. − Gidrobiol. Zh., 3:2. 1967.

Ivanov, A.I. Mussels (Mytilus galloprovincialis Lam.) of the Black Sea and Prospects for their Catch. Author's Summary of Candidate Thesis. Odessa. 1968.

Ivanov, M.V. A Method for Determining the Production of the Bacterial Biomass in Water Bodies. − Mikrobiologiya, 24:1. 1955.

Ivlev, V.S. The Biological Productivity of Water Bodies. − Usp. Sovrem. Biol., 19. 1945.

Ivlev, V.S. The Heterotrophic Region of the Production Process. − Trudy Sevastopol'. Biol. Sta., 15. 1964.

Ivleva, I.V. Growth and Reproduction of Enchytraeus albidus Honle. − Zool. Zh., 32:3. 1953.

Jitts, H.R., C.D. McAllister, K. Stephens, and J.D.H. Strickland. The Cell Division Rates of Some Marine Phytoplankters as a Function of Light and Temperature. − J. Fish. Res. Bd. Can., 21, No.1. 1964.

Johnson, D.F. Growth of Glaucoma ficaria Kahl in Cultures with Single Species of Other Microorganisms. − Arch. Protistenk., 86:3. 1936.

Johnson, M. and J. Olson. The Life History and Biology of a Marine Harpaticoid Copepod Tisbe furcatus (Baird). − Biol. Bull., 95:3. 1948.

Juday, C. Annual Energy Budget of Inland Lake. − Ecology, 21:438. 1940.

Juday, C. The Utilization of Aquatic Food Resources. − Science, 97:2525. 1943.

Kahl, A. Ciliate libera et ectocommensalia. − Tierwelt N.-u. Ostsee, 23. 1933.

Kamshilov, M.M. Production of Calanus finmarchicus (Gunner) in the Littoral Zone of Eastern Murman. − Trudy Murmansk. Morsk. Biol. Sta., 4. 1958.

Karzinkin, G.S. Foundations of the Biologic Productivity of Water Bodies. − Moscow, "Pishcheprom." 1952.

Khailov, K.M. Elements of Ecological Metabolism in the Marine Littoral. Author's Summary of Doctoral Thesis. Moscow. 1969.

Khmeleva, N.N. Energy Expenditures for Respiration, Growth and Reproduction in Artemia salina L. − In: Biologiya morya, 15. Kiev, "Naukova Dumka." 1968.

Kirpichenko, M.Ya. Phenology, Population Dynamics and Growth of Dreissena Larvae in the Kuibyshev Reservoir. − In: Biologiya dreisseny i bor'ba s nei. Moscow−Leningrad, "Nauka." 1964.

Kondrat'eva, T.M. On the Production of Phytoplankton in the Mediterranean Sea. − In: Osnovnye cherty geologicheskogo stroeniya, gidrologicheskogo rezhima i biologii Sredizemnogo morya. Moscow, "Nauka." 1965.

Kondrat'eva, T.M. On the Diurnal Production of Phytoplankton in the Black Sea. − In: Biologicheskie issledovaniya Chernogo morya i ego promyslovykh resursov. Moscow, "Nauka." 1968.

Konstantinov, A.S. On the Method for Determining Production of Animals Consumed by Fish. − Nauch. Dokl. Vyssh. Shk., Biol. Ser., 4. 1960.

Kozhova, O.M. Bacterioplankton of the Irkutsk Reservoir during the First Years after the Filling in 1957−1960. − Trudy Limnol. Inst. Sibirsk. Otdel, Akad. Nauk SSSR, 11, No.31. 1964.

Krasheninnikova, S.A. Microbiological Characteristics of the Gorki Reservoir during the Second Year of its Existence. − Trudy Inst. Biol. Vodokhran., 3, No.6. 1960.

Kruger, F. Stoffwechsel und Wachstum bei Scyphomedusen. − Helgoländer wiss. Meeresunters., 18:4. 1968.

Kryuchkova, N.M. Utilization of Food for Growth in Moina rectirostris Leydig. –
    Zool. Zh., **46**:7. 1967.

Kudelina, E.N. Effect of Temperature on the Reproduction, Development and Fecundity of
    Calanipeda.- Trudy Kasp. filiala Vsesoyuznogo Nauchno-Issledovatel'skogo Instituta
    Morskogo Rybnogo Khozyaistva i Okeanografii (VNIRO), 2. 1950.

Kuenzler, E.J. Structure and Energy Flow of a Mussel Population in a Georgia Salt Marsh. –
    Limnol. Oceanogr., 6:2. 1961.

Kuz'menko, K.N. Life Cycle and Production of Pontoporeia affinis Lindstr. in Lake
    Krasnoe in the Karelian Isthmus. – Gidrobiol. Zh., 5:4. 1969.

Kuznetsov, V.V. Biology and Biological Cycle of Lacuna pallidula Da Costa in the
    Barents Sea. – In: Pamyati akad. Zernova. Moscow–Leningrad, Akad. Nauk SSSR. 1948a.

Kuznetsov, V.V. Bioecological Characteristics of Mass Marine Invertebrate Species. –
    Izv. Akad. Nauk SSSR, Biol. Ser., 5. 1948b.

Kuznetsov, V.V. The White Sea and Biological Features of its Fauna and Flora. – Moscow–
    Leningrad, Akad. Nauk SSSR. 1960.

Kuznetsov, S.I. and V.I. Romanenko. Microbiological Examination of Inland Water Bodies
    (A Laboratory Manual). – Moscow–Leningrad, Akad. Nauk SSSR. 1963.

Lackey, J.B. Occurrence and Distribution of Marine Protozoan Species in the Woods Hole
    Area. – Biol. Bull., 70:26. 1936.

Lanskaya, L.A. Rate and Conditions of Division of Marine Planktonic Algae in Cultures.–
    In: Pervichnaya produktsiya morei i vnutrennikh vod. Minsk. 1961.

Lebedeva, M.N. Growth, Reproduction and Production of Daphnia longispina in the
    Ucha Rezervoir. – Byull. Mosk. Obshch. Ispyt. Prir., otd. biol., 68:5. 1963.

Lie, U. A Quantitative Study of Benthic Fauna in Puget Sound. – FiskDir. Skr., Serie
    Havundersokelser, 14:5. 1968.

Lindeman, R.L. The Trophic-Dynamic Aspect of Ecology. – Ecology, **23**:399. 1942.

Loefer, J.B. Bacteria-Free Culture of Paramecium barsaria and Concentration of the
    Medium as a Factor in Growth. – J. Exp. Zool., 72:3. 1936.

Lohmann, H. Das Gehäuse d. Appendicularien. – Zool. Anz., 22. 1899.

Lohmann, H. Untersuchungen zur Feststellung des vollständigen Gehaltes des Meeres an
    Plankton. – Wiss. Meeresunters., Sec. Kiel, 10:131. 1908.

Lyakhov, S.M. and V.P. Mikheev. Disrtibution and Abundance of Dreissena in the
    Kuibyshev Reservoir during the Seventh Year of its Existence. – In: Biologiya
    dreisseny i bor'ba s nei. Moscow–Leningrad, "Nauka." 1964.

MacFadyen, A. Animal Ecology – Aims and Methods. – London, Pitman. 1957.

MacFadyen, A. Energy Flow in Ecosystems and its Exploitation by Grazing. – In: Grazing
    Terrestr. Marin. Environ. Oxford, Blackwell Sci. Publs. 1964.

Machemer, H. Abhängigkeit der Lebensdauer und Teilung bei Stylonychia mytilus
    von äusseren Faktoren. – Zool. Jahrb., Sec.1, 71:2. 1964.

Makkaveeva, E.V. Biocenosis of Cystoseira barbata in the Coastal Zone of the Black
    Sea. – Trudy Sevastopol'. Biol. Sta., 12. 1959a.

Makkaveeva, E.V. Population and Biomass Dynamics of Rissoa splendida Eichw. off
    the Coasts of Crimea. – Trudy Sevastopol'. Biol. Sta., 11. 1959b.

Makkaveeva, E.V. Small Worms, Crustaceans and Water Mites of the Cystoseria
    Biocenosis. – Trudy Sevastopol'. Biol. Sta., 14. 1961.

Maksimova, L.P. Biology of Moina and Rotifers and their Cultivation as Live Food for the
    Larvae of Whitefishes. – Izv. Gos. Nauchno-Issled. Inst. Ozern. Rechn. Ryb. Khoz.,
    67. 1968.

Malovitskaya, L.M. Population Dynamics of Principal Species of the Sea of Azov Zoo-
    plankton. – In: Biologiya i raspredelenie planktona yuzhnykh morei. Moscow, "Nauka."
    1967.

Margalef, R. Temporal Succession and Heterogeneity in Phytoplankton. – In: Perspectives in Marine Biology, Univ. Calif. Press. 1960.

Margalef, R. Communication of Structure in Planktonic Populations. – Limnol. Oceanogr., 6:2. 1961.

Margalef, R. On Certain Unifying Principles in Ecology. – Am. Nat. , **97**:897. 1963a.

Margalef, R. Role des cilies dans le cycle de la vie pelagique en Mediterranee. – Rapp. P.–v. Réun. Comm. int. Explor. Mer., 17:2. 1963c.

Margalef, R. Nouvelles observations sur la distribution des cilies oligotriches dans le plancton de la Mediterranee Occidentale. – Rapp. P.-v. Réun. Commn. int. Explor. Mer., **19**:3. 1968.

Markosyan, A.K. Biology of Gammarus in Lake Sevan. – Trudy Sevan. Gidrobiol. Sta., 10. 1948.

Matveeva, T.A. Biology of Mytilus edulis L. in Eastern Murman. – Trudy Murmansk. Biol. Sta., 1. 1948.

Matveeva, T.A. Biology and Biological Cycle of Acmaea testudinalis in the Eastern Murman Area. – Trudy Murmansk. Biol. Sta., 2. 1955.

Mironov, G.N. Feeding of Planktonic Predators. – In: Biologiya i raspredelenie planktona yuzhnykh morei. Moscow, "Nauka." 1967.

Mironov, G.N. Linear Growth and Weight Increase of Sagitta setosa Müll. – In: Biologiya morya, 20. Kiev, "Naukova Dumka." 1970.

Moore, H.B. The Biology of Balanus balanoides. – J. Mar. Biol. Ass. U.K., 19:2. 1934.

Mordukhai-Boltovskaya, E.D. Data on the Biology of Infusoria in the Rybinsk Reservoir. – Trudy Inst. Biol. Vnutr. Vod, 8, No.13. 1965.

Morozova-Vodyanitskaya, N.V. and L.A.Lanskaya. Rate and Conditions of Division of Marine Diatom Algae in Cultures. – Trudy Sevastopol'. Biol. Sta., 12. 1959.

Myers, I. Physiology of the Algae. – A. Rev. Microbiol., 5. 1951.

Negus, Ch.L. A Quantitative Study of Growth and Production of Unionid Mussles in the River Thames at Reading. – J.Anim. Exol., 35:3. 1966.

Novozhilova, M.I. Population and Biomass Dynamics of Bacteria in the Water Mass of the Rybinsk Reservoir. – Mikrobiologiya, 24:6. 1955.

Novozhilova, M.I. Generation Time of Bacteria and Production of Bacterial Biomass in the Water of the Rybinsk Reservoir. – Mikrobiologiya, 26:2. 1957.

Oberthur, K. Untersuchungen an Frontonia marina Fabre-Dom. aus einer Binnenland-Salzquelle unter besonderer Berücksichtigung der pulsierenden Vakuole. – Arch. Protistenk., 88:3. 1937.

Odum, E.P. Fundamentals of Ecology. Philadelphia and London, N.B.Saunders Co. 1959.

Osadchikh, V.F. and E.A.Yablonskaya. On the Production of Some Species of the North Caspian Benthos. – In: Metody opredeleniya produktsii vodnykh zhivotnykh. Minsk. 1968.

Paloheimo, J.E. and L.M.Dickie. Food and Growth of Fishes. 1. – J. Fish. Res. Bd. Can., 22:2, 1965.

Parker, R.C. The Effect of Solution in Pedigree Lines of Infusoria. – J. Exp. Zool., 49:2. 1927.

Pavlova, E.V. Developmental Cycle and Some Data on the Growth of Penilia avirostris Dana in Sevastopol Bay. – Trudy Sevastopol'. Biol. Sta., 11. 1959.

Pavlovskaya, T.V. Experimental Study of the Diet of Some Infusorian Species in the Black Sea. – Uspekhi Protozool., Leningrad, "Nauka." 1969.

Pechen',G.A. Production of Cladocerans of Lake Zooplankton. – Gidrobiol. Zh., 1:4. 1965.

Pechen',G.A. and E.A.Shushkina. Production of Planktonic Crustaceans in Different Types of Lakes. – Proc. 10th Conf. on Inland Waters of the Baltic Area. Minsk. 1964.

Petipa, T.S. Average Weight of the Main Forms of the Black Sea Zooplankton. – Trudy Sevastopol'. Biol. Sta., 9. 1956.

Petipa, T.S. Relationship between the Growth Increment, Energy Metabolism and Ration of Acartia clausi Giesbr. – In: Fiziologiya morskikh zhivotnykh. Moscow, "Nauka." 1966a.

Petipa, T.S. On the Energy Balance of Calanus helgolandicus (Claus) in the Black Sea. – Ibid. 1966b.

Petipa, T.S. On the Efficiency of Energy Utilization in Black Sea Pelagic Ecosystems. – In: Struktura i dinamika vodnykh soobshchestv i populyatsii. Kiev, "Naukova Dumka." 1967.

Petrova, M.A. Production of Planktonic Crustaceans in the Gorki Reservoir. – Gidrobiol. Zh., 3:6. 1967.

Petrusevich, K. Basic Concepts in the Study of Secondary Production. – Zh. Obshch. Biol., 28:1. 1967.

Pidgaiko, M.L. Calculation of the Biological Production of Some Cladocerans. – In: Voprosy gidrobiologii. Moscow, "Nauka." 1965.

Pidgaiko, M.L. Potential Production. – In: Metody opredeleniya produktsii vodnykh zhivotnykh. Minsk. 1968.

Poddubnaya, T.L. Life Cycle and Growth Rate of Limnodrilus newaensis Mich., Oligachaeta, Tubificidae. – Trudy Inst. Biol. Vodokhran., 5:8. 1963.

Pütter, A. – Pflügers Arch. ges. Physiol., 180. 1920.

Reeve, M.R. Biology of Chaetognatha, 1. – Symp. Mar. Food Chains, Denmark. Contrib., p.11. 1968.

Richards, O.W. The Correlation of the Amount of Sunlight with the Division Rates of Ciliates. – Biol. Bull., 56:298. 1929.

Richards, O.W. The Growth of Protozoa. – In: Protozoa in Biological Research. New York, Columbia Univ. Press. 1941.

Richards, O.W. and G.A.Riley. The Benthic Epifauna of Long Island Sound. – Bull. Bingham Oceanogr. Coll., 19:2. 1967.

Rodina, A.G. Methods of Aquatic Microbiology. A Practical Manual. – Moscow–Leningrad, "Nauka." 1965.

Romanova, A.P. and A.I. Zonov. Determination of Production of the Bacterial Mass in Water Bodies. – Dokl. Akad. Nauk SSSR, 158:1. 1964.

Salmanov, M.A. Population Dynamics of Bacteria in Waters of the Kuibyshev Reservoir. – Trudy Inst. Biol. Vodokhran., 2, No.5. 1959.

Scherbaum, O and G.Rasch. Cell Size Distribution and Single Cell Growth in Tetrahymens pyriformis Gl. – Acta path. microbiol. scand., 41:3. 1957.

Schuberg, A. Zur Kenntnis des Stentor coezuleus. – Zool. Jahrb.Sec.Anat. Ontog. Tiere, 4. 1891.

Shcherbakov, A.P. Productivity of the Zooplankton of Lake Glubokoe. 3. Planktonic Protozoa. – Trudy Vses. Gidrobiol. Obshch., 13. 1963.

Shcherbakov, A.P. Lake Glubokoe. – Moscow, "Nauka." 1967.

Shelbourne, J.E. A Predator-prey Size Distribution for Plaice Larvae Feeding on Oikopleura.– J. Mar. Biol. Ass. U.K., 42:2. 1962.

Shpet, G.I. Relationship between Size, Occupied Space and Bioproductivity of Some Aquatic Organisms. – Zh. Obshch. Biol., 23:4. 1962.

Shpet, G.I. Procedures for Comparing Productivity of Aquatic Animals. – Proc. 10th Sci. Conf. on Inland Waters of the Baltic Area. Minsk. 1964.

Shpet, G.I. On the Comparative Productivity of Aquatic (and Other) Animals.– Zh. Obshch.Biol., 26:2. 1965.

Shpet, G.I. Biological Productivity of Fishes and Other Animals. – Kiev, "Urozhai." 1968.

Shushkina,E.A. Relationship between the Production and Biomass of Lake Zooplankton. – Gidrobiol. Zh., 2:1. 1968.

Shushkina, E.A. Calculation of the Production of Copepods on the Basis of their Metabolism and the Coefficient of the Utilization of Assimilated Food for Growth. – Okeanologiya, 8:1. 1968.

Shushkina, E.A. and Yu.I. Sorokin. On Determining Production of Zooplankton by the Radiocarbon Method. – Okeanologiya, 9:4. 1969.

Slavina, O.Ya. The Growth of Mussels in Sevastopol Bay. – In: Bentos. Kiev, "Naukova Dumka." 1965.

Slobodkin, L.B. Growth and Regulation of Animal Populations. – Holt, Rinehart, Winston, New York. 1962.

Smidt, E.L.B. Animal Production in the Danish Wadden Sea. – Meddr. Kommn. Danm. Fisk.– Havunders., 11:6. 1951.

Smith, F.E. Quantitative Aspects of Population Growth. – In: Dynamics of Growth Processes. Princeton Univ. Press. 1954.

Sokolova, N.Yu. The Production of Chironomids in the Ucha Reservoir. – In: Metody opredeleniya produktsii vodnykh zhivotnykh. Minsk. 1968.

Sorokin, Yu.I. Some Aspects of the Trophic Role of Bacteria in Water Bodies. – Gidrobiol. Zh., 3:5. 1967.

Stolte, H.A. Morphologische und physiologische Untersuchungen an Blepharisma undulans Stein. – Arch. Protistenk., 48:245. 1924.

Stranghoner, E. Teilungsrate und Kernreorganisationsprozess bei Paramaecium multi-micronucleatum, Powers und Mitchell. – Arch. Protistenk., 78:2. 1932.

Strel'tsov, V.E. Patterns of Postembryonic Growth of the Polychaete Harmathoe imbricata (Polychaeta, Errantia) in the Littoral of the Southern Part of the Barents Sea. – Dokl. Akad. Nauk SSSR, 169:6. 1966.

Stross, R.G., J.C. Nees, and A.D. Hasler. Turnover Time and Production of the Planktonic Crustacea in Lined and Reference Portion of a Bog Lake. – Ecology, 42:237. 1961.

Sushchenya, L.M. Production and Annual Flow of Energy in the Population of Orchestia bottae M.Edw. (Amphipoda – Talitroidea). – In: Struktura i dinamika vodnykh soobshchestv i populyatsii. Kiev, "Naukova Dumka." 1967.

Sushchenya, L.M. and N.N. Khmeleva. Food Consumption as a Function of Body Weight in Crustaceans. – Dokl. Akad. Nauk SSSR, 176:67. 1967.

Taylor, C. Temperature, Growth and Mortality – the Pacific Cockle. – J. Cons. Explor. Mer., 26:117. 1960.

Ten, V.S. Method for Calculating Production of Phytoplankton. – Trudy Sevastopol'. Biol. Sta., 15. 1964.

Ten, V.S. and V.E. Zaika. Main Parameters of the Production Process in Populations of Aquatic Invertebrates. – In: Biologiya i raspredelenie planktona yuzhnykh morei. Moscow, "Nauka." 1967.

Thienemann, A. Der Produktionbegriff in der Biologie. – Archiv. Hydrobiol., 22:4. 1931.

Tsikhon-Lukanina, E.A. Nutrition and Growth of Freshwater Gastropod Mollusks. – Trudy Inst. Biol. Vnutr. Vod, 9. 1965 a.

Tsikhon-Lukanina, E.A. Life Cycle, Population and Biomass Dynamics of Some Gastropod Mollusks in the Littoral Zone of the Rybinsk Reservoir during the Summer of 1962. – Trudy Inst. Biol. Vnutr. Vod, 8. 1965 b.

Vasil'eva, G.L. Cultivation of Brachionus rubens Ehrbg. as Food for Fish Larvae. – Gidrobiol. Zh., 4:5. 1968.

Vernadskii, V.I. Chemical Structure of the Earth's Biosphere and its Surroundings. – Moscow, "Nauka." 1965.

Vevers, H.C. The Biology of Asterias rubens L.: Growth and Reproduction. – J. Mar. Biol. Ass. U.K., 28, No.1. 1949.

Vinberg, G.G. Some General Problems of the Productivity of Lakes. – Zool. Zh., 15:4. 1936.

V i n b e r g, G.G. Metabolic Rate and Food Requirement of Fishes. Minsk. 1956.

V i n b e r g, G.G. The Energy Principle in the Study of the Trophic Links and Productivity of Ecologic Systems. – Zool. Zh., 41:11. 1962.

V i n b e r g, G.G. The Biotic Balance of Matter and Energy and the Biological Productivity of Water Bodies. – Gidrobiol. Zh., 1:1. 1965.

V i n b e r g, G.G. The Growth and Metabolic Rates of Animals. – Usp. Sovrem. Biol., 6. 1966.

V i n b e r g, G.G. (editor). Methods for Determining the Production of Aquatic Animals. – Minsk, "Vysheishaya Shkola." 1968 a.

V i n b e r g, G.G. General Aspects of Animal Growth. – In: Metody opredeleniya produktsii vodnykh zhivotnykh. Minsk, "Vysheishaya Shkola." 1968 b.

V i n b e r g, G.G. Interrelationship of Metabolic and Growth Rates of Animals. – In: Biologiya morya, 15. Kiev, "Naukova Dumka." 1968 c.

V i n b e r g, G.G., G.A. P e c h e n ', and E.A. S h u s h k i n a. Production of Planktonic Crustaceans in Three Lakes of Different Types. – Zool. Zh., 44:5. 1965.

V o d y a n i t s k i i, V.A. On the Biological Productivity of Water Bodies with Particular Reference to the Black Sea. – Trudy Sevastopol'. Biol. Sta., 8. 1954.

V o r o b' e v, V.P. Mussels of the Black Sea. – Trudy Azov-Chernomorsk. Nauchno-Issled. Inst. Morsk. Ryb. Khoz. Okeanogr., 11. 1938.

V o r o b' e v, V.P. The Sea of Azov Benthos. – Trudy Azov-Chernomorsk. Nauchno-Issled. Inst. Morsk. Ryb. Khoz. Okeanogr., 13. 1949.

W e s t p h a l, A. Studien über Ophryoscoleciden in der Kultur. – Z. Parasit Kde, 7:1. 1934.

W e y e r, G. Untersuchungen über die Morphologie und Physiologie des Formwechsels der Gastrostyla steinii Engelmann. – Arch. Protistenk., 71:1. 1930.

W h i t s o n, G.L. Temperature Sensitivity and its Relation to Changes in Growth, Control of Cell Division and Stability of Morphogenesis in P a r a m e c i u m   s u r e l i a Syngen Stock 51. – J. Cell. Comp. Physiol., 64:3. 1964.

W i c h t e r m a n, R. Descriptions and Life Cycle of E u p l o t e s   n e a p o l i t a n u s Sp. nov. (Protozoa, Giliophora, Hypotrichida) from the Gulf of Naples. – Trans. Am. Microsc. Soc., 83:3. 1964.

W i l l i a m s, R.B. Division Rates of Salt Marsh Diatoms in Relation to Salinity and Cell Size. – Ecology, 45:1. 1964.

W o o d, E.J.F. Marine Microbial Ecology. – London–New York Reinholds Publs. 1965.

W o o d r u f f, L.L. P a r a m e c i u m   a u r e l i a in Pedigree Culture for Twenty-Five Years. – Trans. Am. Microsc. Soc., 51:3. 1932.

W o o d r u f f, L.L. and G.A. B a i t s e l l. The Temperature Coefficient of the Rate of Reproduction of P a r a m e c i u m   a u r e l i a. – Am. J. Physiol., 29:2. 1911.

W o o d r u f f, L.L. and E.L. M o o r e. On the Longevity of Ŝ p a t h i d i u m   s p a t h u l a without Endomixis or Conjugation. – Proc. Natn. Acad. Sci. U.S.A., 10:5. 1924.

W o o d r u f f, L.L. and H. S p e n c e r. Studies on S p a t h i d i u m   s p a t h u l a. – J. Exp. Zool., 39:2. 1924.

W r i g h t, J.C. The Population Dynamics and Production of D a p h n i a in Canyon Ferry Reservoir, Montana. – Limnol. Oceanogr., 10:4. 1965.

Y a b l o n s k a y a, E.A. Attempts to Use E.V. Borutskii's Method for Determining the Production of Chironomids. – In: Metody opredeleniya produktsii vodnykh zhivotnykh. Minsk. 1968.

Z a i k a, V.E. On the Biology and Production of Appendicularians in the Black Sea. – In: Voprosy morskoi biologii. Kiev, "Naukova Dumka." 1966.

Z a i k a, V.E. Research Objects and Applicability of Some Concepts in Synecology. – In: Struktura i dinamika vodnykh soobshchestv i populyatsii. Kiev, "Naukova Dumka." 1967 a.

Z a i k a, V.E. Methods for the Calculation of the Production of Bacteria. – Okeanologiya, 7:3. 1967 b.

Z a i k a, V.E. Age-Structure Dependence of the "Specific Production" in Zooplankton
    Populations. – Mar. Biol. 1:311. 1968.

Z a i k a, V.E. On the Production of Appendicularians and Sagittas in the Neritic Zone of the
    Black Sea. – In: Produktsionno-biologicheskie protsessy v planktone yuzhnykh morei.
    Kiev, "Naukova Dumka." 1969 a.

Z a i k a, V.E. Results of the Study of Planktonic Infusorians in the Mediterranean. – In:
    Ekspeditsionnye issledovaniya v Sredizemnom more v 1968 g. Kiev, "Naukova
    Dumka." 1969 b.

Z a i k a, V.E. Reproduction Rates of Infusorians. – In: Produktsiya i pishchevye svyazi v
    soobshchestvakh planktonnykh organizmov. Kiev, "Naukova Dumka." 1970 a.

Z a i k a, V.E. Productivity of Aquatic Mollusks as a Function of the Life Span. – Okeanologiya,
    10:4. 1970 b.

Z a i k a, V.E. Rapports entre la productivité des Mollusques Aquatiques et la durée de leur vie. –
    Cah. Biol. mar., 11:99. 1970 c.

Z a i k a, V.E. Relationship between the Maximal Specific Growth Rates of Warm-blooded
    Animals. – Zool. Zh., 49:2. 1970 d.

Z a i k a, V.E. The Specific Production of Aquatic Invertebrates. Doctoral Thesis. Moscow. 1970 e.

Z a i k a, V.E. and A.A. A n d r y u s h c h e n k o. Specific Production as a Function of Age Structure
    of Zooplankton Populations. – Zh. Obshch. Biol., 30:3. 1969.

Z a i k a, V.E. and T.Yu. A v e r i n a. Density of Infusoria in the Plankton of Sevastopol Bay in the
    Black Sea. – Okeanologiya, 8:6. 1968.

Z a i k a, V.E. and T.Yu. A v e r i n a. Division Rates of Some Species of Black Sea Infusoria. –
    In press.

Z a i k a, V.E. and N.P. M a k a r o v a. The "Parabolic Growth" and Coefficient of the Utilization
    of Assimilated Food for Growth. – In: Voprosy morskoi biologii. Kiev, "Naukova
    Dumka." 1969.

Z a i k a, V.E. and N.A. M a k a r o v a. Theoretical Analysis of the Production Process in the
    Bacterioplankton. – In: Produktsiya i pishchevye svyazi v soobshchestvakh planktonnykh
    organizmov. Kiev, "Naukova Dumka." 1970.

Z a i k a, V.E. and N.P. M a k a r o v a. Biological Significance of the Parameters of the Bertalanffy
    Growth Equation. – Dokl. Akad. Nauk SSSR, 199. 1971 a.

Z a i k a, V.E. and N.P. M a k a r o v a. On the Possible Unity of the Growth Potentialities of
    Organisms. – Zool. Zh., 50:3. 1971 b.

Z a i k a, V.E. and L.M. M a l o v i t s k a y a. Characteristics of the Variability of the Specific
    Productivity of Some Zooplanktonic Populations. – In: Struktura i dinamika vodnykh
    soobshchestv i populyatsii. Kiev, "Naukova Dumka." 1967.

Z a i k a, V.E. and N.A. O s t r o v s k a y a. Growth Rate, Life Span and Specific Production in
    Mollusks. – Zh. Obshch. Biol., 32:3. 1971.

Z a i k a, V.E., E.V. P a v l o v a, and A.V. K o v a l e v. Diet of Planktonic Crustaceans in the
    Mediterranean Sea.– In: Ekspeditsionnye issledovaniya v Sredizemnom more v 1968 g.
    Kiev, "Naukova Dumka." 1969.

Z a i k a, V.E. and T.V. P a v l o v s k a y a. Feeding of Marine Infusorians on Unicellular Algae. –
    In: Produktsiya i pishchevye svyazi v soobshchestvakh organizmov. Kiev, "Naukova
    Dumka." 1970.

Z e n k e v i c h, L.A. Data on Diet of Fishes in the Barents Sea. – Proc. of the 1st Session of the
    Oceanogr. Inst., 4. 1931.

Z e n k e v i c h, L.A. Productivity of USSR Seas. – Proc. Faunistic Conf. Zool. Inst., Sec.
    Hydrobiol. Leningrad. 1934.

Z e n k e v i c h, L.A. Fauna and Biological Productivity of the Sea, 1:134. 1947.

Z e n k e v i c h, L.A. Fauna and Biological Productivity of the Sea, 2. 1951.

Z h d a n o v a, G.A. Comparative Study of the Life Cycle and Productivity of B o s m i n a l o n g i-
    r o s t i s and B . c o r e g o n i Baird in the Kiev Reservoir. – Gidrobiol. Zh., 5, No.1. 1969.

SUBJECT INDEX